HACCP: AN INTEGRATED APPROACH TO ASSURING THE MICROBIOLOGICAL SAFETY OF MEAT AND POULTRY

F
N
P
PUBLICATIONS IN
FOOD SCIENCE AND NUTRITION

Journals

JOURNAL OF FOOD LIPIDS, F. Shahidi

JOURNAL OF RAPID METHODS AND AUTOMATION IN MICROBIOLOGY,
D.Y.C. Fung and M.C. Goldschmidt

JOURNAL OF MUSCLE FOODS, N.G. Marriott, G.J. Flick, Jr. and J.R. Claus

JOURNAL OF SENSORY STUDIES, M.C. Gacula, Jr.

JOURNAL OF FOODSERVICE SYSTEMS, C.A. Sawyer

JOURNAL OF FOOD BIOCHEMISTRY, J.R. Whitaker, N.F. Haard and H. Swaisgood

JOURNAL OF FOOD PROCESS ENGINEERING, D.R. Heldman and R.P. Singh

JOURNAL OF FOOD PROCESSING AND PRESERVATION, D.B. Lund

JOURNAL OF FOOD QUALITY, J.J. Powers

JOURNAL OF FOOD SAFETY, T.J. Montville

JOURNAL OF TEXTURE STUDIES, M.C. Bourne and M.A. Rao

Books

OF MICROBES AND MOLECULES: FOOD TECHNOLOGY AT M.I.T., S.A. Goldblith

MEAT PRESERVATION: PREVENTING LOSSES AND ASSURING SAFETY,
R.G. Cassens

S.C. PRESCOTT, M.I.T. DEAN AND PIONEER FOOD TECHNOLOGIST,
S.A. Goldblith

FOOD CONCEPTS AND PRODUCTS: JUST-IN-TIME DEVELOPMENT, H.R. Moskowitz

MICROWAVE FOODS: NEW PRODUCT DEVELOPMENT, R.V. Decareau

DESIGN AND ANALYSIS OF SENSORY OPTIMIZATION, M.C. Gacula, Jr.

NUTRIENT ADDITIONS TO FOOD, J.C. Bauernfeind and P.A. Lachance

NITRITE-CURED MEAT, R.G. Cassens

POTENTIAL FOR NUTRITIONAL MODULATION OF AGING, D.K. Ingram *et al.*

CONTROLLED/MODIFIED ATMOSPHERE/VACUUM PACKAGING OF
FOODS, A.L. Brody

NUTRITIONAL STATUS ASSESSMENT OF THE INDIVIDUAL, G.E. Livingston

QUALITY ASSURANCE OF FOODS, J.E. Stauffer

THE SCIENCE OF MEAT AND MEAT PRODUCTS, 3RD ED., J.F. Price and
B.S. Schweigert

HANDBOOK OF FOOD COLORANT PATENTS, F.J. Francis

ROLE OF CHEMISTRY IN PROCESSED FOODS, O.R. Fennema *et al.*

NEW DIRECTIONS FOR PRODUCT TESTING OF FOODS, H.R. Moskowitz

PRODUCT TESTING AND SENSORY EVALUATION OF FOODS, H.R. Moskowitz

ENVIRONMENTAL ASPECTS OF CANCER: ROLE OF FOODS, E.L. Wynder *et al.*

FOOD PRODUCT DEVELOPMENT AND DIETARY GUIDELINES, G.E. Livingston, R.J.
Moshy, and C.M. Chang

SHELF-LIFE DATING OF FOODS, T.P. Labuza

ANTINUTRIENTS AND NATURAL TOXICANTS IN FOOD, R.L. Ory

UTILIZATION OF PROTEIN RESOURCES, D.W. Stanley *et al.*

VITAMIN B_6: METABOLISM AND ROLE IN GROWTH, G.P. Tryfiates

POSTHARVEST BIOLOGY AND BIOTECHNOLOGY, H.O. Hultin and M. Milner

Newsletters

MICROWAVES AND FOOD, R.V. Decareau

FOOD INDUSTRY REPORT, G.C. Melson

FOOD, NUTRITION AND HEALTH, P.A. Lachance and M.C. Fisher

FOOD PACKAGING AND LABELING, S. Sacharow

HACCP: AN INTEGRATED APPROACH TO ASSURING THE MICROBIOLOGICAL SAFETY OF MEAT AND POULTRY

Edited by

James J. Sheridan, Ph.D.

Head, Meat Technology Department
Teagasc, The National Food Centre
Dunsinea, Castleknock
Dublin 15, Ireland

Robert L. Buchanan, Ph.D.

Research Leader
Microbial Food Safety Research Unit
United States Department of Agriculture
600 East Mermaid Lane
Philadelphia, Pennsylvania

Thomas J. Montville, Ph.D.

Professor
Rutgers University
Department of Food Science
Cook College
New Brunswick, New Jersey

FOOD & NUTRITION PRESS, INC.
TRUMBULL, CONNECTICUT 06611 USA

CONTRIBUTORS

ROBERT L. BUCHANAN, Research Leader, Microbial Food Safety Research Unit, United States Department of Agriculture, 600 East Mermaid Lane, Philadelphia, PA 19118, USA.

JOHN D. COLLINS, Associate Professor of Veterinary Preventive Medicine, Chair of Department of Large Animal Clinical Studies, Faculty of Veterinary Medicine, University College, Dublin, Ireland.

CHARLES DALY, Professor, Faculty of Food Science and Technology, University College, Cork, Ireland.

MICHAEL P. DOYLE, Professor, Head, Department of Food Science and Technology, The University of Georgia, Centre for Food Safety and Quality Enhancement, Georgia Station, Griffin, GA 30223-1797, USA.

BIN FU, Research Scientist, Frito Lay, Inc., 7701 Legacy Drive, Plano, TX 75024, USA.

CHARLES P. GERBA, Professor of Microbiology and Soil and Environmental Sciences, University of Arizona, Tucson, AZ 85721, USA.

DALE M. GROTELUESCHEN, Associate Professor, Director of Veterinary Extension and Diagnostics, University of Nebraska, and Director of Panhandle Research and Extension Center, Scottsbluff, NE 69361, USA.

CHARLES N. HAAS, Betz Professor of Environmental Engineering, Environmental Studies Institute, Department of Civil and Architectural Engineering, Drexel University, Philadelphia, PA 19104, USA.

THEODORE P. LABUZA, Professor of Food Science and Technology, Department of Food Science and Nutrition, University of Minnesota, St. Paul, MN 55108, USA.

MAJELLA MAHER, Research Officer, National Diagnostics Centre, BioResearch Ireland, University College, Galway, Ireland.

ANN M. McNAMARA, Director, Microbiology Division, Food Safety and Inspection Service, United States Department of Agriculture, Washington, DC 20250, USA.

JIANGHONG MENG, Post-doctoral Associate, The University of Georgia, Centre for Food Safety and Quality Enhancement, Georgia Station, Griffin, GA 30223-1797, USA.

GERALDINE MOLLOY, Research Officer, National Diagnostics Centre, BioResearch Ireland, University College, Galway, Ireland.

DAN MONTANARI, Director, Management Information Systems, Coleman Natural Meats, Inc., 5140 Race Court 4, Denver, CO 80216, USA.

ROEL W.A.W. MULDER, Manager, International Research Affairs, DLO-Institute for Animal Science and Health, Agricultural Research Department (DLO-NL), PO Box 59, 6700 AB Wageningen, The Netherlands.

JOAN B. ROSE, Assistant Professor, Department of Marine Sciences, University of South Florida, St. Petersburg, FL 33701, USA.

SHARIN SACHS, Associate Director, Information and Legislative Affairs, USDA Food Safety and Inspection Service, Room 1175, South Building, Washington, DC 20250, USA.

JAMES J. SHERIDAN, Head, Meat Technology Department, Teagasc, The National Food Centre, Dunsinea, Castleknock, Dublin 15, Ireland.

GREGORY R. SIRAGUSA, Microbiologist, United States Meat Animal Research Center, United States Department of Agriculture, Agricultural Research Service, PO Box 166, Clay Center, NE 68933, USA.

TERRY SMITH, Senior Research Scientist, National Diagnostics Centre, BioResearch Ireland, University College, Galway, Ireland.

PAUL SOCKETT, Chief of Division of Disease Surveillance, Bureau of Infectious Diseases, Laboratory Centre for Disease Control, Ottawa, Ontario, Canada K1A OL2.

PATRICK V. TARRANT, Director of Operations, Teagasc, The National Food Centre, Dunsinea, Castleknock, Dublin 15, Ireland.

DAN W. THAYER, Research Leader, Food Safety Research Unit, U.S. Department of Agriculture, Agricultural Research Service, Eastern Regional Research Center, 600 East Mermaid Lane, Philadelphia, PA 19118, USA.

MARY UPTON, Lecturer in Food Microbiology, Department of Industrial Microbiology, University College, Dublin, Belfield, Dublin 4, Ireland.

SHAOHUA ZHAO, Post-doctoral Associate, The University of Georgia, Centre for Food Safety and Quality Enhancement, Georgia Station, Griffin, GA 30223-1797, USA.

TONG ZHAO, Agricultural Research Coordinator, The University of Georgia, Centre for Food Safety and Quality Enhancement, Georgia Station, Griffin, GA 30223-1797, USA.

PREFACE

The microbiological safety of the food supply, particularly meat, has become an increasingly important concern to the public, and is a factor affecting international trade. One of the systems for the control of foodborne pathogens that is being recommended is the implementation of hazard analysis of critical control points (HACCP). This is a practical approach to ensuring the safety and quality of meat and poultry which is being adopted by the food industry worldwide. The application of HACCP also has the approval of government institutions in the United States and within the European Union which are involved in the development and implementation of safety controls within the food industry. Both industry and governments view HACCP as the most practicable and cost effective way of controlling food safety.

This publication contains papers presented at a conference held in Dublin on March 23-24, 1994. The objective of this conference was two-fold. The first was to provide a "farm to fork" overview of how HACCP techniques can be used throughout the entire food production-manufacturing-distribution-consumption chain to help control foodborne pathogens of public concern in meat and poultry. The second was to use these presentations as the basis for identifying the knowledge and research that will be needed in the future to use HACCP to develop practical food safety systems. The program was presented under four main headings: (1) farm practices; (2) slaughter practices; (3) post-slaughter practices, and (4) consumer aspects. The conference was organized by Dr. J.J. Sheridan of The National Food Centre, Teagasc, Ireland and Dr. R.L. Buchanan of the USDA, Eastern Regional Research Laboratory and was opened by the Minister for Agriculture Food and Forestry, Mr. Joe Walsh, T.D. The keynote addresses were delivered by Dr. P. Power for the Irish side and by Dr. Jill Hollingsworth on behalf of the agencies of the USDA.

In recognition of five years of progress under the US/Ireland Cooperation in Agricultural Science & Technology Programme, this international conference was sponsored by:

Ireland: Department of Agriculture, Food and Forestry

USA: The following agencies of the United States Department of Agriculture:

Office of International Cooperation and Development
Food Safety and Inspection Service
Cooperative State Research Service
Agricultural Research Service

J.J. SHERIDAN
R.L. BUCHANAN
T.J. MONTVILLE
August 1995

CONTENTS

Session III. Post Slaughter Operations
Chairman: Professor Charles Daly

Session IV. Consumer Aspects
Chairman: Dr. P.V. Tarrant

IMPACT OF TRANSPORT AND RELATED STRESSES ON THE INCIDENCE AND EXTENT OF HUMAN PATHOGENS IN PIGMEAT AND POULTRY

R.W.A.W. MULDER

DLO-Institute for Animal Science and Health
(The former Spelderholt Centre for Poultry Research and Information Services)
Agricultural Research Department (DLO-NL)
P.O. Box 59, 6700 AB Wageningen, The Netherlands

ABSTRACT

Meat products are very important sources of protein in the human diet. The contamination of these products with pathogenic microorganisms, such as Salmonella *and* Campylobacter, *make both production and consumption of them a precarious proposition. Several methods can decrease the level of contamination with these pathogenic microorganisms. However, there are still situations of high microbial load which cannot be explained and are often attributed to stress. This paper describes the effect of transport, husbandry and nutrition practices on contamination of slaughtered products. Examples of the stress occurring during fattening, catching and loading, transport and conditioning at the processing plant, are given.*

INTRODUCTION

Meat forms an important component in the Western diet. Over the years, there have been increasing requirements from consumers for foods that are safe, do not need the addition of preservatives and need little or no preparation or cooking by the consumer. Thus there is a real pressure to produce meat and meat products that contain minimal numbers of both spoilage and human pathogenic organisms.

Human foodborne diseases are considered to be one of the major problems in the modern world, and they are an important cause of economic losses due to hospitalization and absenteeism. *Salmonella* and *Campylobacter* infections account for the majority of acute cases of human gastroenteritis. The results of sentinel and population studies (Table 1), carried out in The Netherlands since 1987, demonstrate that *Campylobacter* bacteria are the most prominent cause of acute gastroenteritis in humans. This seems in contrast to traditional reports,

1

where *Salmonella* is the top-seeded organism. This difference is probably caused by the past (and in some cases continued use) use of investigating and reporting systems, that do not include *Campylobacter*.

   Salmonella spp., *Campylobacter* spp., *Listeria monocytogenes* and *Staphylococcus aureus* are the potentially pathogenic microorganisms which are most frequently isolated from live animals and pig and poultry meat.

TABLE 1.
CAUSES AND INCIDENCES OF ACUTE GASTROENTERITIS IN MAN
IN THE NETHERLANDS 1987-1992

Microorganisms	Frequency (%)	Incidence (#/1000 individuals per year)
Campylobacter	12–15	18–23
Salmonella	4–5	6–11
Shigella	0–3	
Escherichia coli	3	
Clostridium perfringens	3	
Rotaviruses	6	

Adapted from Notermans and van de Giessen (1993).

   Although the reports of isolations are not consistent for all organisms from all commodities, some trends can be observed. *Salmonella* isolations are most frequently reported; the most important *Salmonella* serotypes in isolates from humans, poultry and pigs are shown in Table 2. Unfortunately, comparable data are not available for *Campylobacter* and other pathogenic microorganisms. From these results, it is evident that even in a two year period the top-five *Salmonella* serotypes have changed. In 1985, *S. enteritidis* accounted for approx 2.4% of the human isolates in The Netherlands. In 1989 this figure was 20.1% and rose to 34.4% of the isolates in 1991. The data in Table 2 also suggest that *S. typhimurium* infections in humans are caused by pigs, and *S. enteritidis* infection by poultry.

   In a recent survey, Jacobs-Reitsma *et al.* (1994), estimated the *Campylobacter* and *Salmonella* incidence in broiler flocks over a one year period from early 1992 to early 1993. *Campylobacter* spp. were isolated from 153 out of 187 broiler flocks (82%). *Campylobacter jejuni* was the dominant species, although due to novel serotyping systems the difference with *C. coli* isolates is not very clear. There was a seasonal variation with maximum isolation rates in June-September and minimum isolations in March. In the same flocks the *Salmonella* contamination was not influenced by the season and *Salmonella* was isolated from 49 out of 181 flocks (27%). Consumer-ready poultry products were *Campylobacter*-positive in 62.6% of the samples and *Salmonella*-positive in 44.2% of the total number of samples (Bolder 1993).

TABLE 2.
TOP FIVE *SALMONELLA* SEROTYPES IN ISOLATES FROM HUMANS,
POULTRY AND PIGS IN THE NETHERLANDS

	Human (%)		Poultry (%)		Pig (%)	
	1989	1990	1989	1990	1989	1990
S. typhimurium	44.9	39.7	16.9	18.4	81.2	77.6
S. enteritidis	20.1	29.5	19.5	10.0	1.0	<1
S. virchow	6.2	6.8	8.6	13.1	<1	0
S. hadar	2.2	2.6	9.9	18.8	0	<1
S. infantis	2.4	1.9	11.3	14.2	1.9	1.5

Data: National Institute of Public Health and Environmental Protection, Bilthoven, The Netherlands

In contrast to poultry, where *Campylobacter jejuni* is the predominant species, pigs are believed to be the most important source of *Campylobacter coli*. Faecal carriage rates of *Campylobacter* spp. among pigs may be as, in poultry, up to 100%. Contamination of slaughter pigs before chilling is up to 30%, after overnight chilling this is below 3%. The treatment of pigs before slaughter influences contamination of pig products with these organisms. The length of stay in lairage is an important factor in reducing *Salmonella* contamination of pig products (Morgan *et al.* 1987).

The reduction of contamination of live animals with *Salmonella* and other potentially pathogenic microorganisms has been the subject of study in many countries over the last 25 years. Until now no single treatment or process able to eradicate *Salmonella* and other pathogens could be identified. Even the mechanism of infection with these organisms could not completely be elucidated for all potentially pathogenic microorganisms. The use of competitive exclusion has been somewhat successful. The treatment of day-old broiler chicks with a microbiota which colonizes them and confers resistance to colonization by pathogens, the so-called competitive exclusion treatment, reduces the contamination of flocks by *Salmonella* and *Campylobacter* under laboratory and farm conditions. In studies where flocks were also transported to the slaughterhouse and examined again for these microorganisms, higher contamination rates were found (Goren *et al.* 1988). Similar results were reported for pigs. The collection of the animals on the farm, their transport to the processing plant and the holding time and conditions before slaughtering seem to induce the spreading of organisms, resulting in higher contamination and a continued carrier state of the animals.

Salmonella and other pathogenic microorganisms may be transferred from one animal to another when they are waiting slaughter, via faeces and drinking troughs. The prevention of an excessively long period of holding in lairages and prevention of overcrowding, specially in pigs, will considerably reduce the proportion of animals found contaminated at slaughter. Clinically healthy animals carrying *Salmonella* and other pathogenic microorganisms may change their excretion pattern of the organisms from intermittent to constant shedding if an external factor upsets the equilibrium of their intestinal flora. A disturbance of the intestinal functions will lower the resistance of the live animal and facilitates the spreading of intestinal bacteria.

"Stress" is the word often used in those situations which are too complex to be understood, and it is often used to explain why preventive measures to control spreading of pathogens in live animals do not work. Stress factors described in the literature were observed during fattening, catching and loading, transport and conditioning. Stress can, among others, be accompanied by symptoms such as damage to the intestinal tract and a lower capacity of the immune system.

The influence of stress on the immune system is complex and depends on a number of factors. Among these are: the stressor, genetics, nutrition, antigen concentration. Some stressors are believed to influence positively the resistance to infections with pathogens. Examples are some forms of so-called social stress which seem to increase resistance against *S. aureus* and *E. coli*. On the other hand, other social stressors, for example mixing pigs from different herds together at transport, have resulted in higher rates of contamination in pigs. (Williams and Newell 1968; Gallwey and Tarrant 1979). Renwick *et al.* 1993, recently, demonstrated, that products became more contaminated with microorganisms when the time between crating and holding before slaughter increased. This indicates shedding of faecal material, which spreads over live birds. The number of hours of feed withdrawal prior to crating also influences excretion of pathogens. Normally chickens empty their caeca every 24 h, but because of the change in environmental conditions the excretion pattern changes. In the literature most data relate to spreading/shedding of *Salmonella* bacteria. Although *Campylobacter* seem to cause more problems with regard to human public health, this prominent position is not reflected yet in the literature.

EXAMPLES FROM PIGS AND POULTRY

Pigs

Preslaughter handling can affect the contamination rate of slaughtered animals. Feeding, environmental conditions during transport and lairage, including the total time involved and mixing animals from several herds, are the main factors. Slavkov *et al.* (1974) demonstrated the effect of stress during loading, transport and holding time before slaughter. Before pigs were delivered to the abattoir, none were isolated; after delivering to the abattoir 0.1% (2/1952) of the samples were positive; after slaughter this percentage increased to 0.7%. The authors conclude, that stress factors had been responsible for the increase in the carrier state.

The most comprehensive work in this area is by Morgan *et al.* (1987). These authors studied the effect of time spent in lairage on caecal and carcass *Salmonella* contamination of groups of pigs originating from one producer and slaughtered in two different abattoirs. The main differences between the abattoirs were pen size and (visual) hygiene (abattoir 1: larger pens and less hygiene). Table 3 presents some data from this study. Thus the time spent in lairage can be used to minimize *Salmonella* contamination. The shorter the period, the better. Pen size (smaller pens) and hygiene are the other important factors to decrease (cross)contamination. In this study carcass contamination was caused by intestinal *Salmonella* infections. This could be demonstrated by the *Salmonella* recovery rate and the *Salmonella* serotypes from caecal contents and the carcass surface.

TABLE 3.
ISOLATION OF *SALMONELLA* FROM CAECAL CONTENTS OF PIGS HELD IN
LAIRAGE FOR THREE DAYS (18, 42 AND 66 H) IN TWO ABATTOIRS
Number of positives/total number sampled.

| Abattoir | Caecal isolation rate | | |
	Day 1	Day 2	Day 3
1	20/75	29/71	45/75
2	8/76	6/74	24/74
Total	28/151	35/145	71/149
	(18.5%)	(24.1%)	(47.7%)

Adapted from Morgan *et al.* (1987).

As carcass contamination was determined by the *Salmonella* entering the abattoir in the intestine of the pigs, a very important strategy to reduce contamination is preslaughter handling avoiding any form of multiplication of *Salmonella* in the live animals (see also Huis in't Veld *et al.* 1994). Another reason for spending a short time in lairage, is the economic aspect of the carcass weight loss with increasing time.

Poultry

Stress in poultry is accompanied by a series of symptoms. The increased corticosteroid levels in blood plasma and the occurrence of damage to the intestinal tract, heart and blood vessels are of major importance. Decreased shear strength of the intestinal tract may result in gut breakage during processing, which is responsible for further spreading of microorganisms over carcasses and equipment (Bilgili 1988). Feed and water withdrawal prior to transport, influences gut contents and the emptying of the digestive tract of broilers. Papa and Dickens (1989) concluded that feed withdrawal 8–12 h before slaughtering minimizes the faecal contamination of carcasses. Moran and Bilgili (1990) demonstrated that stressing chicken broilers, under conditions simulating the practice of feed withdrawal and live haul, results in a delayed caecal retention for another 24 h.

Bolder and Mulder (1983) reported the increase of *Salmonella* contamination of slaughtered broilers after transport. The similarity with findings in the pig industry, as in other areas, is striking: *Salmonella* serotypes after transport observed on slaughtered products originate from live birds, which indicates intestinal origin. The question arises whether there is not a major discrepancy between husbandry and nutrition factors aiming at the economical production of poultry broilers and the contamination of consumer-ready products with potentially pathogenic microorganisms. Preslaughter conditions influence the contamination rate of slaughtered products. However, the real circumstances and mechanisms are not known. To interrupt microbial cycles in animal production, more attention should be paid to aspects of contamination in relation to husbandry, nutrition and processing, including conditions of loading and transport.

CONCLUSIONS

The mechanism of spreading/shedding of microorganisms from clinically healthy carrier animals under stressing conditions is not clear. Nevertheless from the literature and actual practice, it is known that preslaughter conditions in handling live animals influence the contamination rate of the slaughtered product. Therefore preloading activities in the shed, the loading procedure, the transport, the holding period and conditions at slaughter and the slaughter process itself should be given more care.

REFERENCES

BILGILI, S.F. 1988. Effect of feed and water withdrawal on shear strength of broiler gastro-intestinal tract. Poultry Sci. *67*, 845–847.

BOLDER, N.M. 1993. *Salmonella, Campylobacter, Listeria* and *Escherichia coli* O157:H7 monitoring in Dutch poultry products on retail level. Spelderholt Confidential Rept. 695–15.

BOLDER, N.M. and MULDER, R.W.A.W. 1983. Contamination des carcasses de poulets pas des Salmonelles: Le role des caisses de transport. Courr. Avicole *39*, 23–25.

GALLWEY, W.J. and TARRANT, P.V. 1979. Influence of environmental and genetic factors on the ultimate pH in commercial and pure-bred pigs. Acta Agric. Scandinavica Suppl. *21*, 32–38.

GOREN, E., DE JONG., W.A., DOORNENBAL, P., BOLDER, N.M., MULDER, R.W.A.W. and JANSEN, A. 1988. Reduction of *Salmonella* infection in broilers by spray application of intestinal microflora: a longitudinal study. Vet. Q. *10*, 249–255.

HUIS IN'T VELD, J.H.J., MULDER, R.W.A.W. and SNIJDERS, J.M.A. 1994. Impact of animal husbandry and slaughter technologies on microbial contamination of meat: monitoring and control. Meat Sci. *36*, 123–154.

JACOBS-REITSMA, W.F., BOLDER, N.M. and MULDER, R.W.A.W. 1994. *Campylobacter* and *Salmonella* in broiler flocks. Intestinal carriage of *Campylobacter* and *Salmonella* in Dutch broiler flocks at slaughter: a one-year study. Poultry Sci. *73*, 1260–1266.

MAY, J.D. and DEATON, J.W. 1989. Digestive tract of broilers cooped or deprived of water. Poultry Sci. *68*, 627–630.

MORAN, JR., E.T. and BILGILI, S.F. 1990. Influence of feeding and fasting market age broilers on cecal access to an oral dose of *Salmonella*. J. Food Prot. *53*, 205–207.

MORGAN, J.R., KRAUTIL, F.L. and CRAVEN, J.A. 1987. Effect of time lairage on caecal and carcass *Salmonella* contamination of slaughter pigs. Epidemiol. Infect. *98*, 323–330.

NOTERMANS, S. and VAN DE GIESSEN, A. 1993. Foodborne diseases in the 1980s and 1990s. Food Control *4*, 122–124.

PAPA, C.M. and DICKENS, J.A. 1989. Lower gut contents and defecatory responses of broiler chickens as affected by feed withdrawal and electrical treatment at slaughter. Poultry Sci. *68*, 1478–1484.

RENWICK, S.A., McNAB, W.B., LOWMAN, H.R. and CLARKE, R.C. 1993. Variability and determinants of carcass bacterial load at a poultry slaughter abattoir. J. Food Prot. *56*, 694–699.

SIEGEL, H.S. 1987. Effects of behavioural and physical stressors on immune response. In *Biology of Stress in Farm Animals: An Integrative Approach,* (P.R. Wiepkema and P.W.M. van Adrichem, eds.), pp. 39–55, Martinus Nijhof Publishers.

SLAVKOV, I., IODANOV, I., MILEV, M. and DANOV, V. 1974. Study of factors capable of increasing the number of *Salmonella* carriers in clinically normal pigs before slaughter. Veterinarnomeditsinski Nauki, Bulgaria, *11*, 88–91.

WILLIAMS, L.P. and NEWELL, K.W. 1968. Sources of salmonellae in market swine. J. Hyg. *66*, 281–293.

CHAPTER 2

IMPACT OF FARM MANAGEMENT PRACTICES ON THE INCIDENCE OF HUMAN ENTERIC PATHOGENS IN CATTLE

DALE M. GROTELUESCHEN

University of Nebraska
Panhandle Research & Extension Center
Scottsbluff, Nebraska

ABSTRACT

Food safety is of vital importance to the world meat industry. United States livestock producers have implemented voluntary pre-harvest commodity-based programs to address quality related issues, such as microbial and chemical contamination. Examples include the Beef Quality Assurance Program (National Cattlemen's Association), Pork Quality Assurance Program (National Pork Producers Council), and the Dairy Quality Assurance Program (National Milk Producers Federation).

Improvement of microbiological control at production levels is valuable in increasing food safety to consumers, in improving animal health, productivity and economic return to producers and in decreasing risk to those working in the livestock industry. Human enteric pathogens, such as Salmonella, *cause significant health and economic losses in animal production. Practising veterinarians have implemented control and prevention programs. Information is needed for risk assessment and development of methods to reduce risk from pathogens of concern.*

Employment of hazard analysis of critical control points (HACCP) principles has been received positively by the livestock industry. Great diversity within the industry has made the application of HACCP principles difficult. For example, there are more than 900,000 cow/calf operations in the United States, ranging in size from a few head to thousands of animals. The Food Animal Production Medicine Consortium has recently proposed a HACCP model for livestock production to enhance animal health.

The need for new research to advance efforts to improve both human and animal health is stressed.

INTRODUCTION

Microbiological food safety is an important issue for beef in the food chain. Minimization of risk from exposure to potential human pathogens on beef is a goal of the entire industry. Cattle producers are implementers of management practices to reduce risk and are supportive of research for improvement.

Cattle production in the United States involves a very diverse industry. Fed beef originates from over 900,000 cow/calf producers distributed throughout the country, while mature beef comes from cow/calf producers and from over 200,000 dairy producers. Typically, calves are weaned by cow/calf producers at about 180-318 kg (400-700 lb.). A portion of these are kept for further feeding on the same premises before movement to a feedlot. Others are marketed to buyers called backgrounders, who feed the calves for a period of time before movement to a feedlot. An increasing number of calves are transported directly to feedlots for feeding and reach finished market weight at 14-16 months. Over 45,000 feedlots are responsible for the finishing phase of fed cattle production.

Health Programs

Potential for microbial contamination exists for both fed and mature beef at production stages. Management practices are very different between livestock operations but many of the same risks are shared. Herd health programs promoting improved animal well-being are very common in cow/calf and feedlot operations. These programs typically give attention to management practices that reduce stress and exposure for animals. Vaccinations also play major roles. Nutritionists are employed to ensure that adequate rations are fed. Immune competence is an important factor in disease-prevention strategies.

The beef cattle industry in the United States is committed to providing safe, wholesome beef to consumers. The industry is market driven. Management practices which improve animal health and potential for profit also tend to reduce risk of exposure and infection with agents that are also pathogenic to humans.

Health Risks

There are aspects of beef production that can influence the health of animals and also the risk of exposure to infection with potential human pathogens. The production cycle typically involves time spent on two or more premises prior to movement to processing facilities. The transportation required contributes to both stress placed upon the animals and risk of exposure to potential human pathogens (Cole *et al.* 1988; Hutcheson and Cole 1986). Dairy animals handled in the same way experience similar risks.

The marketing phase can also increase pathogen exposure, especially through mingling of animals and through pathogen build-up at marketing facilities. Animals are marketed typically through auction markets, direct selling from cow/calf producer to backgrounder or feedlot, video auctions, and collection points. Fed cattle are most often sold directly to processing facilities.

Effects of Stress

Stressed animals have lowered resistance to disease, so that they are more susceptible to pathogens and at increased risk of shedding potential human pathogens (Breazile 1988). For example, stressed animals exposed to *Salmonella* would be at greater risk of becoming infected and subsequent shedders of this organism. Various stresses can result in increased levels of shedding, infection of pen-mates, clinical disease, and economic loss to the owner. In addition there is an increased risk to humans from direct animal contact and from microbial contamination of meat.

Stress reduction through improved production systems that result in better animal health would be expected to reduce levels of pathogen shedding and the incidence of potential human pathogens in cattle production systems. For example, improvements in handling systems reduce stress-related immune suppression associated with animal processing procedures (Grandin 1984; 1987). Immune-suppressed animals can be expected to be more severely affected by other stress and exposure factors, such as animal density, frequency of feedlot pen use, and co-mingling of sick animals. *Salmonella* is capable of surviving for prolonged periods in animal environments (Rings 1985).

Feed and Water

Feed and water are potential sources of microbial contamination of cattle (Robinson *et al.* 1992). Feedstuffs should be free of enteric pathogens, especially *Salmonella* (Mitchell and McChesney 1991).

Quality Assurance

The United States beef industry has in recent years implemented programs designed to improve the quality of beef for consumers. Beef quality assurance programs were begun in the late 1980s to draw attention to the importance of quality in the entire production phase and its effect on the marketed product. These programs promote improved animal health, better management, and increased awareness of consumer needs. Recently, Total Quality Management (TQM) programs have provided leadership about continuous improvement in the beef cattle industry, giving the consumer top priority. To summarise, the beef industry has become increasingly consumer driven, and food safety is a primary emphasis.

Industry Cooperation

Increased vertical cooperation between segments of the industry has great potential for improvement of cattle health. Recently a Strategic Alliances Field Study (SAFS), managed by the National Cattlemen's Association in coordination with Colorado State and Texas A & M Universities has been completed. This has demonstrated how selected management practices and animal health programs, combined with consistent communication, resulted in highly positive outcomes in a number of important parameters, including health (Anon. 1993). This study involved 1,253 calves from 15 different ranch operations located in five states. Ranches were required to meet certain management and vaccination criteria designed to reduce stress and also to enhance immunological status upon entry into a single selected feedlot. Calves were purchased by a partnership consisting of the feedlot, a packing company, and the original ranch owners. Calves were transported directly to and fed in the feedlot, where a number of parameters were measured, both during the feeding period, at processing, and at the retail level. The results related to health are discussed here.

SAFS cattle were compared to groups of Ranch-Fresh (RF) calves (4,193) and Put-Together (PT) calves (1,830). The RF calves were shipped directly from the premises of cow/calf producers with no preventative management or treatment. The PT calves were from numerous sources which were co-mingled upon entry into the feedlot. Calves which required veterinary treatment from these groups, expressed as a percentage of the total were: SAFS, 22.4; RF, 33.0, and PT 22.7. Of the calves which recovered some subsequently required further treatment as follows; SAFS, 3.8%; RF, 21.5%, and PT, 16.0%. Treatment costs for the respective groups were as follows; SAFS, $1.92, RF, $6.29, and PT, $4.95. Percent death losses from each group were; SAFS, 1.2, RF, 1.81, and PT, 2.95.

The results of the health component of this study have shown that the calves receiving management practices and preventative measures, known to improve health performance in the feedlot, had a lower sick re-treatment rate, lower mortality, and lower treatment costs than the other groups. These animals required less antibiotic use and fewer injections during the feeding period. Management practices are available to improve health risks. If potential human pathogen risk is also lowered using systems such as this, the risk of microbial contamination can also be reduced using practices that are beneficial to the animals, the cow/calf producer, the feedlot, and most importantly, the consumer.

Recommendations for Food Safety

The United States Food Animal Production Medicine Consortium recently sponsored a workshop entitled: Implementing Food Animal Pre-Harvest Food Safety Internationally. One outcome of this was a ten-point summary on implementing food animal pre-harvest (pre-slaughter) food safety internationally (Anon. 1992). A brief synopsis is as follows:

(1) Pre-harvest food safety is important and essential.
(2) Food safety requires global perspective.
(3) Voluntary cooperation and participation is possible and should succeed at all levels.
(4) Pre-harvest food safety issues require excellent science, education, and skill in many disciplines.
(5) Pre-harvest food safety is likely to be an important "value added" issue and must be economically sound.
(6) New educational curricula are needed for professionals and producers.
(7) Pre-harvest food safety research is essential. Examples: on-farm practices, economic output, environment, animal welfare.
(8) Public health, production, and economic goals must be met and assessment is needed.
(9) An international team approach is needed.
(10) Pre-harvest food safety is central to other issues such as economic development, sustainable agriculture, environment and resources, and welfare of animals.

SUMMARY

Beef production is a large, dynamic industry in the United States. The industry is sensitive to consumer needs and is addressing important issues. Effects of management on animal health at the production level is being addressed with known information. Immunology, especially immuno-suppressive effects of some practices, is an important aspect, and influences risk of infection with and shedding of potential human pathogens. Additional research is needed to further reduce the risk of exposure of cattle to human pathogens. This would include: identification of management practices directly associated with decreased risk of human pathogen exposure; the identification of high risk animal production systems; better tests to identify carrier animals and herds. Methods to eradicate potential human pathogens from carrier animals and herds, including prevention of re-infection are also important.

The beef industry is adopting continuous improvement principles that allow complex issues such as microbial food safety to be effectively addressed.

REFERENCES

ANON. 1992. Food animal production medicine consortium. Implementing food animal pre-harvest food safety internationally. *Proceedings: Providing Safe Food for the Consumer.* Food Animal Production Medicine Consortium. Washington, DC.

ANON. 1993. Strategic alliances field study. National Cattlemen's Assoc., Colorado State University, and Texas A & M University, 2-7.

BREAZILE, J.E. 1988. The physiology of stress and its relationship to mechanisms of disease and therapeutics. *The Veterinary Clinics of North America: Food Animal Practice*. Vol. *4*, (3) 441-480. W.B. Saunders Co., Philadelphia.

COLE, N.A., CAMP, T.H., ROWE JR., L.D., STEVENS, D.G. and HUTCHESON, D.P. 1988. Effect of transport on feeder calves. Amer. J. Vet. Res. *49*, 178-183.

GRANDIN, T. 1984. Reduce stress of handling to improve productivity of livestock. Vet. Med. *79*, 827-831.

GRANDIN, T. 1987. Using cattle psychology to aid handling. Agri Practice *8* (5): 32-36.

HUTCHESON, D.P. and Cole, N.A. 1986. Management of transit-stress syndrome in cattle: Nutritional and environmental effects. J. Anim. Sci. *62*, 550-560.

MITCHELL, G.A. and McCHESNEY, D.G. 1991. A plan for *Salmonella* control in animal feeds. *Proceedings of Symposium on the Diagnosis and Control of Salmonella*. pp. 28-31, U.S. Animal Health Assoc., San Diego, CA.

RINGS, D.M. 1985. Salmonellosis in calves. *The Veterinary Clinics of North America: Food Animal Practice*. Vol. *1*, (3) 529-539. W.B. Saunders Co., Philadelphia.

ROBINSON, R.A., FERRIS, K.E., MILLER, D.A. and SRINAND, S. 1992. Descriptive epidemiology of *Salmonella* serotypes from cattle in the USA (1982-1991). *XVII World Buiatrics Congress. Amer. Assoc. of Bovine Practitioners Conference*, pp. 15-19, St. Paul, MN.

CHAPTER 3

USE OF VACCINE AND BIOLOGICAL CONTROL TECHNIQUES TO CONTROL PATHOGENS IN ANIMALS USED FOR FOOD

SHAOHUA ZHAO, JIANGHONG MENG, TONG ZHAO and
MICHAEL P. DOYLE

*Center for Food Safety and Quality Enhancement and
Department of Food Science and Technology
University of Georgia
Georgia Station
Griffin, Georgia, USA*

ABSTRACT

An important component of a Hazard Analysis Critical Control Point (HACCP) approach applied to animal production is reducing the carriage of food-associated pathogens by animals. Two approaches that have both great potential for reducing pathogen colonization of animals and merit for practical application include (1) vaccination and (2) competitive exclusion. Vaccination involves acquisition of immunity in an animal following exposure to an attenuated pathogen or an antigen of a virulent microorganism. Adherence factors that influence bacterial colonization of animals are useful antigens for vaccines. A strategy to developing a vaccine to reduce carriage of pathogens includes (1) identifying sites of colonization by the pathogen in the animal, (2) defining the mechanism of colonization, (3) characterizing genes that encode colonization factors, (4) transforming the colonization factor antigen genes into a suitable nonpathogenic vector, and (5) determining the optimal methods of immunization. Studies are underway to identify colonization factors of Escherichia coli O157:H7 for use in vaccine development to protect cattle from colonization by this pathogen. Competitive exclusion involves the use of microbial cultures that out-compete pathogens from colonizing specific niches. A science-based approach to identifying/developing useful competitive exclusion cultures is to: (1) define how a pathogen colonizes the site of interest, (2) isolate microbes that colonize the same site and produce metabolites that inhibit or kill the pathogen, and (3) verify that the inhibitory microbes, when introduced into pathogen-free animals, can reduce or prevent colonization by the pathogen.

This approach has been used successfully to identify defined bacterial cultures that can greatly reduce colonization of Campylobacter jejuni in poultry. Commercial implementation of techniques of these types is an essential part of the overall HACCP approach to reducing the prevalence of foodborne pathogens from farm to fork.

INTRODUCTION

Animals used for food frequently carry in their intestinal tracts, human pathogens (Doyle 1989; Schoeni and Doyle 1992). *Campylobacter jejuni* and *Salmonella* colonize the intestinal tracts of poultry, swine, sheep and cattle, generally without ill effect (Doyle 1989; Schoeni and Doyle 1992). Enterohemorrhagic *Escherichia coli* O157:H7 has been isolated in many surveys from faeces of cattle (Griffin and Tauxe 1991; Wells *et al.* 1991, Zhao *et al.* 1995). Faecal contamination of hides, feathers, hair and external surfaces of animals is common and to be expected considering the living habits of farm-raised animals. Hence pathogens in faeces also contaminate these surfaces. Removal of the hides or feathers during slaughter operations can result in pathogen contamination of carcasses and subsequent contamination of meat. Therefore, reducing the carriage of pathogens by animals entering the food chain would directly reduce the chance contamination that may result in human disease.

Vaccination to Prevent/Reduce Carriage of Pathogens

Two approaches that can be practically implemented and have great potential for reducing pathogen colonization of animals are vaccination and competitive exclusion. Vaccination involves eliciting an immune response to a pathogen in an animal by introducing an attenuated pathogen, an avirulent microorganism closely related antigenically to the pathogen, or an antigen of a virulent microorganism. Generally, a successful vaccine is able to provide strong and long-lasting immunity, is nontoxic to the host, is easy to administer and store, and is inexpensive to produce. A recent innovation in this area is the application of recombinant DNA technology to develop subunit recombinant vaccines which have great potential in meeting the above criteria.

Depending on the objective of vaccination, the type of vaccine and the administration procedure, host immune responses are different. In some situations, humoral and mucosal immune responses are principally involved with protection of animals, whereas in other cases cell mediated immunity is more important. For enteric pathogens, gut mucosal immunity is particularly important because gut mucosal specific secretory immunoglobulin (sIgA) can serve as a first line of defense by directly interacting with invading pathogens and preventing pathogen colonization and invasion of the gastrointestinal tract. Orally administered vaccines, especially live attenuated vaccines, are effective in inducing specific IgA responses, presumably because antigen is delivered to T and B lymphocytes of gut-associated lymphoid tissue. Primed B cells then migrate to mesenteric lymph nodes and undergo differentiation. These B cells enter the thoracic duct and then the general circulation, subsequently seeding secretory tissues of the body, including the lamina propria of the gut. IgA is then produced by mature plasma cells and is transported onto the mucosal surface where it is available to interact with invading pathogens.

A strategy to developing a vaccine to reduce carriage of pathogens includes: (1) identifying sites of colonization of the pathogen in the animal, (2) defining the mechanism of colonization, (3) characterizing the genes that encode colonization factors, (4) transforming the colonization factor antigen genes into a suitable non-pathogenic vector, and (5) determining the optimal methods of immunization. Identifying the sites and mechanisms of colonization is useful for vaccine design and evaluation. Such information is needed to identify colonization factors whose gene sequences can be identified and inserted into an appropriate vector. This construct is likely to be a suitable vaccine candidate because host antibodies should be elicited to the colonization factor antigens, thereby preventing the pathogen from establishing itself in its niche within the host. Studies also will be needed to identify the best immunization strategy for producing the greatest immune response.

Vaccine Development for *E. coli* O157:H7

Present knowledge of the pathogenicity of *E. coli* O157:H7 indicates that bacteria adhere to the intestinal epithelium of the host by an attachment and effacing mechanism and subsequently produce one or more Shiga-like toxins (Griffin and Tauxe 1991; Padhye and Doyle 1992). By the host producing antibodies to *E. coli* O157:H7 in its intestinal tract the pathogen will be less able to colonize the gut. Identifying the *E. coli* O157:H7 colonization factor genes and cloning them into a suitable vaccine vector could yield a useful subunit recombinant vaccine for *E. coli* O157:H7.

Identification of colonization factors of *E. coli* O157:H7. Studies using transposon (Tn*phoA*) mutagenesis revealed that outer membrane proteins of *E. coli* O157:H7 may be adherence/colonization factors (Dytoc *et al.* 1993). In addition, the attachment and effacing (*eae*) gene of *E. coli* O157:H7 has recently been identified in a homologous study of the *eae* gene of enteropathogenic *E. coli* O157:H7 (Beebakhee *et al.* 1992; Yu and Kaper 1992). This gene is thought to encode an attachment factor; however, the gene product(s) and its function have not been well characterized.

The Tn*pho*A mutagenesis system has been used for identification of colonization factors of *E. coli* O157:H7. This system allows the creation of mutants deficient in bacterial factors that normally are secreted or exported to the bacterial cell surface (Hoiseth and Stocker 1981). Because virtually all bacteria protein implicated as colonization factor are extracellular, surface-associated, or piroplasmic, the application of Tn*pho*A provides a good opportunity for mutations that affect the colonization properties of bacteria. Several OMPs of *E. coli* O157:H7 have been associated with colonization by Tn*pho*A mutagenesis studies (Dytoc *et al.* 1993).

Cloning and Expressing the Genes in Non-Pathogenic E. coli. Genes encoding colonization factors can be isolated from chromosomal DNA of Tn*pho*A mutants of *E. coli* O157:H7. Restriction endonucleases such as *Eco*R V and M*lu* I that do not cut within Tn*pho*A can be used to generate Tn*pho*A fragments that contain the target genes. Subsequently, the Tn*pho*A fragments can be cloned into plasmid vectors and the target genes can be sequenced by chain termination reaction. A complete gene can be obtained by PCR based on the sequence data. The gene can subsequently be cloned into a plasmid vector and expressed in nonpathogenic *E. coli* using standard methods (Sambrook *et al.* 1989).

Competitive Exclusion to Prevent/Reduce Carriage of Pathogens. Competitive exclusion involves the use of microbial cultures that outcompete pathogens from colonizing specific niches. This approach was first applied successfully by Rantala and Nurmi (1973) to reduce infection of *Salmonella infantis* in chickens. Feeding chicks intestinal bacteria from pathogen-free adult chickens before exposure to *Salmonella* protected most chicks from colonization by salmonellae.

A science-based approach to identifying/developing useful competitive exclusion cultures is to: (1) identify the location(s) of colonization by the pathogen, (2) define how the pathogen colonizes the animal, (3) isolate microbes that colonize the same site(s) and produce metabolites that inhibit or kill the pathogen, and (4) verify that the inhibitory microbes, when introduced into pathogen-free animals, can reduce or prevent colonization by the pathogen. An example of how this approach has been used successfully is the development/identification of defined competitive exclusion cultures to reduce carriage of *Campylobacter jejuni* by poultry (Schoeni and Doyle 1992). An overview of this approach and the results follow.

Identification of Location of Colonization by C. jejuni. Initial studies focused on the sites of colonization by *C. jejuni*. Beery *et al.* (1988) determined that *C. jejuni* preferentially localizes in the chick's lower gastrointestinal tract, especially in the ceca where campylobacters are often detected in populations of 10^4 to 10^7 cells per gram. Histological examination of ceca revealed large populations of densely-packed campylobacters in the lumen of mucus-filled crypts (Beery *et al.* 1988; Doyle 1991). Examination of the crypts by electron microscopy revealed that the campylobacters were present throughout the lumina but were not in direct contact with microvilli (Doyle 1991). Hence, campylobacters appear to colonize crypt mucus without attaching or adhering to crypt surfaces. Similar observations were made by Lee *et al.* (1986) who studied the colonization of mice by *C. jejuni*. They determined that campylobacters colonize mucus deep within intestinal crypts, preferentially colonizing cecal crypts.

Mechanism of Intestinal Colonization by C. jejuni. If campylobacters colonize mucus but do not adhere to the surface of crypts as observed in mice by Lee *et al.* (1986) and as determined in chicks by Beery *et al.* (1988), then by what mechanism are campylobacters attracted to and retained in the mucus? Chemotaxis, which involves the movement of an organism towards or away from a chemical stimulus, can be an important factor in the association of pathogens with mammalian hosts (Freter *et al.* 1979). Hugdahl *et al.* (1989) studied the chemotactic behavior of *C. jejuni* and determined that campylobacters are chemoattracted to mucin, a high molecular weight glycoprotein that is the principal constituent of mucus. Mucin is largely composed of carbohydrate molecules and L-fucose is a major constituent. Hugdahl *et al.* (1989) assayed more than twenty carbohydrates, including constituents of sialic acid that are the major carbohydrates present in mucin, and identified L-fucose as the only chemoattractant for *C. jejuni*.

In addition to its chemoattractant activity, mucin can be metabolized as a sole substrate for growth of *C. jejuni* (Beery *et al.* 1988; Hugdahl *et al.* 1988). Considering these observations the following mechanism of *C. jejuni* colonization of chicks was proposed (Beery *et al.* 1988; Doyle 1991). The site of colonization is the lower gastrointestinal tract where campylobacters localize predominantly in cecal and cloacal crypts. Chemoattraction of *C. jejuni* to mucin attracts the bacteria to mucus, in which it moves by a highly active flagellum to mucus-filled crypts where the organism localizes. *C. jejuni* then grows on mucin as a sole substrate. Campylobacters are likely to remain in the crypt because of their attraction to and metabolism of mucin. *C. jejuni* does not appear to adhere to the intestinal tract suggesting that attachment is not important for colonization of the ceca and cloaca. This information is useful for developing strategies to identify appropriate protective bacteria to reduce/prevent colonization of *C. jejuni*.

Isolate Microbes That Colonize Same Niche as C. jejuni, and Produce Anti-C. jejuni Metabolites. Theoretically microbes that could be used effectively as competitive exclusion cultures would occupy the same niche as *C. jejuni* and produce metabolites that would inhibit or kill campylobacters. Those bacteria that would occupy the same niche would likely possess similar colonization properties as *C. jejuni*, e.g., chemoattracted to mucin and able to utilize mucin as a sole substrate. Hence, studies were done to isolate cecum-colonizing bacteria with the above characteristics from hens that were not colonized by *C. jejuni* (Schoeni and Doyle 1992). Four bacterial strains were isolated from hens' ceca that produced anti-*C. jejuni* metabolites and could utilize mucin as a sole substrate.

Verify the Effectiveness of Anti-C. jejuni Microbes in Reducing/Preventing Colonization by the Pathogen. A mixture of three strains of the cecal colonizing, anti-*C. jejuni* metabolite-producing bacteria were fed to chicks to determine if these bacteria could reduce *C. jejuni* colonization of ceca. Four feeding trials revealed the three strain mixture provided 43 to 100% (average 78%) protection from *C. jejuni* colonization. The dominant cecal bacterium of chicks treated with the three-strain mixture was consistently an *E. coli* O13:H⁻ strain. Three feeding trials using only *E. coli* O13:H⁻ as the protective bacteria resulted in 49 to 72% (average, 59%) protection from *C. jejuni* colonization. These studies verified that the protective bacteria could greatly reduce carriage of *C. jejuni* by poultry.

Commercial implementation of techniques such as vaccination or competitive exclusion whereby carriage of pathogens by animals used in food production is reduced, is an essential part of the total HACCP approach to reducing the prevalence of foodborne pathogens from farm to fork.

REFERENCES

BEEBAKHEE, G., LOUIS, M., DEAZAVEDO, J. and BRUNTON, J. 1992. Cloning and nucleotide sequence of the *eae* gene homologue from enterohemorrhagic *Escherichia coli*. FEMS Microbiol. Lett. *91*, 63–68.

BEERY, J.T., HUGDAHL, M.B. and DOYLE, M.P. 1988. Colonization of gastrointestinal tracts of chicks by *Campylobacter jejuni*. Appl. Environ. Microbiol. *54*, 2365–2370.

DOYLE, M.P. (ed.) 1989. Foodborne Bacterial Pathogens, Marcel-Dekker, New York.

DOYLE, M.P. 1991. Colonization of chicks by *Campylobacter jejuni*. In *Colonization Control of Human Bacterial Enteropathogens in Poultry*. (L.C. Blankenship, ed.), Academic Press, San Diego.

DYTOC, M., SONI, R., CODKERILL, F., DeAZAVEDO, J., LOUIE, M., BRUNTON, M. and SHERMAN, P. 1993. Multiple determinants of verotoxin-producing *Escherichia coli* O157:H7 attaching-effacement. Infect. Immun. *61*, 3382-3391.

FRETER, R., O'BRIEN, P.C.M. and MACSAI, M.S. 1979. Effect of chemotaxis on the interaction of cholera vibrios with intestinal mucosa. Am. J. Clin. Nutr. *32*, 128-132.

GRIFFIN, P.M. and TAUXE, R.V. 1991. The epidemiology of infections caused by *Escherichia coli* O157:H7, other enterohemorrhagic *E. coli*, and the associated hemolytic uremic syndrome. Epidemiol. Rev. *13*, 60-98.

HOISETH, S.K. and STOCKER, B.A.D. 1981. Aromatic-dependent *Salmonella typhimurium* are non-virulent and effective as live vaccines. Nature *291*, 238-239.

HUGDAHL, M.B., BEERY, J.T. and DOYLE, M.P. 1988. Chemotactic behavior of *Campylobacter jejuni*. Infect. Immun. *56*, 1560-1566.

LEE, A., O'ROURKE, J.L., BARRINGTON, P.J. and TRUST, T.J. 1986. Mucus colonization as a determinant of pathogenicity in intestinal infection by *Campylobacter jejuni*: a mouse cecal model. Infect. Immun. *51*, 536-546.

PADHYE, N.V. and DOYLE, M.P. 1992. *Escherichia coli* O157:H7: epidemiology, pathogenesis, and methods for detection in food. J. Food Prot. *55*, 555-565.

RANTALA, M. and NURIMI, E. 1973. Prevention of the growth of *Salmonella infantis* in chick by the flora of the alimentary tract of chickens. Br. Poultr. Sci. *14*, 627-630.

SAMBROOK, J., FRITSCH, F.F. and MARRIATIS, T. 1989. *Molecular Cloning*, 2nd ed. Cold Spring Harbor Laboratory Press, Cold Spring Harbor, New York.

SCHOENI, J.L. and DOYLE, M.P. 1992. Reduction of *Campylobacter jejuni* colonization of chicks by cecum-colonizing bacteria producing anti-*C. jejuni* metabolites. Appl. Environ. Microbiol. *58*, 664-670.

WELLS, J.G., SHIPMAN, L.D. and GREENE, K.D. 1991. Isolation of *Escherichia coli* of serotype O157:H7 and other Shiga-like toxin-producing *E. coli* from cattle. J. Clin. Microbiol. *29*, 985-989.

YU, J. and KAPER J.B. 1992. Cloning and characterization of the *eae* gene of enterohemorrhagic *Escherichia coli* O157:H7. Mol. Microbiol. *6*, 411-417.

ZHAO, T., DOYLE, M.P., SHERE, J. and GARBER, L. 1995. Prevalence of enterohemorrhagic *Escherichia coli* O157:H7 in a survey of dairy herds. Appl. Environ. Microbiol. *61*, 1290-1293.

CHAPTER 4

THE ROLE OF ANIMAL IDENTIFICATION SYSTEMS IN CONTROLLING THE SPREAD OF PATHOGENS

DAN MONTANARI

Coleman Natural Meats, Inc.
5140 Race Court 4, Denver, CO 80216

ABSTRACT

No HACCP program can be fully integrated until the identification and tracking of an animal, and its resultant meat products, can be achieved. As ownership of animals and meat change during the production process, valuable information is gathered, but is seldom shared. Through the aid of modern computers and electronic identification, a system capable of monitoring animals and their meat products can improve food safety. This system can identify the origin, ownership, premises, production methods, feeds, medications, and labelling claims. Once a problem is found, the meat from an individual animal can be recalled, even if the meat has been sold to multiple entities. A HACCP system which includes individual animal identification and recall abilities can improve food safety assurances, while providing a single source of identification for integrating procedures at all phases of production. The use of an identification system, in relation to HACCP is discussed.

INTRODUCTION

For the past 45 years, the agricultural industry has met the challenge of producing low cost food efficiently. However, in obtaining this goal, certain environmental and safety concerns were overlooked. Whether real or perceived, consumers the world over are losing confidence in the food industry's ability to address these concerns.

Recent media coverage of beef burger contamination with *Escherichia coli* in the USA has caused concern. Most consumers have little understanding of the complexity of the food industry, and its attempts to resolve these concerns. As more food products are shipped internationally, their safety is imperative. Questionable production practices have resulted in some countries banning the importation of certain food items, in order to protect their food supply system. Without a method for verifying production practices, many quality products are kept from potential international markets.

Recent hearings held by the US Department of Agriculture (USDA) have resulted in demand for a tracking system that can provide trace back capabilities.

Recently the US Secretary of Agriculture has pledged to introduce a bill empowering the USDA to adopt a system to trace tainted meat back to its source.

STATEMENT OF PURPOSE

Through the use of a single identification number that is applied to an animal, Origen$^{(TM)}$ is able to maintain a reference to this number through to the finished product. As a result, the same number that was once attached on the animal's ear tag is also found on the resulting pieces of meat produced from that animal. Origen$^{(TM)}$ is adaptable to future technologies. It is designed to be used as an information and food safety resource.

Origen$^{(TM)}$ coordinates information on live animals to their resulting food products. It transcends identification barriers between ranches, feed yards, packing plants, processing plants and retailers. Individual pathogen tests, Hazard Analysis at Critical Control Points (HACCP), and other procedures are coordinated with the identification number that is attached to each animal. Labelling claims are verified through to the resulting information record that Origen$^{(TM)}$ maintains.

REQUIREMENTS OF A SYSTEM

Farm to Table

Currently two schools of thought are developing relative to animal identification. One believes that tracking an animal, say 60 days prior to slaughter, is sufficient to create an effective trace back system. The second recognizes that a trace back system will never be complete until it is able to go from the consumer back to its point of origin. This ability to trace a product from the consumer back to its point of origin will be referred to as "Farm to Table".

The information requirements to track an animal 60 days prior to slaughter are quite different from being able to trace an animal from "Farm to Table". In the case of meat, tracking an animal 60 days prior to slaughter simply requires that an identification number be assigned to each animal, which references the premises prior to slaughter. However, this form of identification is still unable to reference the animal's point of origin.

By attaching an identification number to an animal at the place of birth, no additional costs are incurred. The information record now begins at birth and will provide valuable information from that time forward. The only additional resource required is the ability to transfer the animal's identification record when ownership of the animal changes.

By extending animal identification back to the point of origin, an animal can be traced from birth to slaughter. However, the requirements for a system of tracking from "Farm to Table" insist that the identification process proceed forward with the animal after slaughter and remain through all facets of production and distribution.

The considerations for a "Farm to Table" identification system demand not only a reference to the point of origin, but also identification through to the finished product. The only reasonable way to achieve such a system is to identify and track both the animal and its resultant meat products.

Objections

The thought of attempting to identify every meat animal world-wide may cause some heads to spin. Trying to take into account all of the meat industry's identification needs could potentially become a nightmare. However, with the aid of computers and electronic identification, such a system becomes manageable. It might be argued that an identification system will be a burden to small producers. However, many small producers are presently identifying individual animals. Furthermore, the system is presently being used by ranchers of all sizes who raise cattle for a Colorado meat company.

Others may argue that the cost of such a system will be enormous. Many meat companies that employ an inventory control system have already demonstrated a willingness to identify and track their inventories. An animal identification system will simply be an enhancement to their current systems.

Others may question the need for a new system of identification. Current systems of trace back are achieved by a manual, time-consuming, search through records and files. Once these records are found by a single segment, additional efforts must be expended in order to integrate these records with that of other producers. Such a process does not inspire a great deal of confidence or timeliness.

Since many of the current safety and inspection methods were developed prior to the widespread use of computers and bar-codes, no standards currently exist for a form of electronic identification. A new specification for electronic identification will modernize current practices and allow for information integration among parties.

UNIFIED IDENTIFICATION NUMBER

Considerations

The Origen$^{(TM)}$ system of animal identification is cost effective and easy to use. It is adaptable for use from the smallest animal producer to the largest meat packer. The identification system that Origen$^{(TM)}$ employs is adaptable to future requirements and serves a multitude of purposes. In its simplest form it is able to identify ownership and in its most complex, it can trace products from the consumer back to the point of origin.

The system is designed with foresight toward free trade and global markets. Its value far exceeds its cost and effort to implement. This identification system is adaptable to existing industry systems and is not an additional form of regulation. In short, there are good reasons for businesses to adopt this process.

Advantages

The Origen$^{(TM)}$ animal identification allows for the creation of a dynamic data resource. It transcends individual producers and provides valuable information to help research promote higher quality products, as well as safer food. Another advantage is its ability to track pathogens and animal diseases back to their source. Animals raised in one country can be tracked through to market in another country. Also, with the increase of foreign trade agreements, the possibility of animal diseases entering a country will become less of a concern with this system. Current HACCP methods can be greatly improved by a system that can track a product back to its point of origin.

Adaptability

The numbering scheme used by the Origen$^{(TM)}$ system is adaptable for use by large and small producers. The identification number, while possessing certain standards, also possesses the ability to be used for different purposes. The Origen$^{(TM)}$ numbering system can track products based upon lot identity, premises identity, or specific unit identity. In other words, Origen$^{(TM)}$ utilizes a single numbering scheme which can be used in several different ways, while still maintaining compatibility with the other two schemes.

The Origen$^{(TM)}$ system reflects real world situations. Occasionally identification devices are lost or damaged. As a result, the system is able to begin identification at any segment in the production process. Even though the problem of lost identification occurs, every measure is taken to ensure that the problem is kept to a minimum.

Specifications

The specification for the Origen$^{(TM)}$ numbering scheme consists of two sequences similar to that used for UPC codes. The first sequence identifies the point of origin. The second sequence is designed to be used as a reference number. This reference number takes one of the following three forms: (1) a lot number; (2) a premises identity number; (3) specific unit identification number.

The basis of the system is an alpha-numeric identification scheme. One constraint in the case of animals, is that it must be able to fit on an ear tag. In the future, advances in stacked bar-code technology may render this constraint obsolete. However, the most cost effective form of animal identification today is a bar-coded ear tag.

By using an alpha-numeric scheme, rather than a simple numeric sequence, certain advantages are achieved. The following demonstrates the differences between a base 10 and base 36 scheme.

Number of digits	Base 10	Base 36
4 digits:	9,999	1,679,616
5 digits:	99,999	60,466,176
6 digits:	999,999	2,176,782,336

By using a based 36 scheme, a 6 digit prefix will identify over 2 billion producers, and a 6 digit suffix will identify over 2 billion individual items per producer. Of course more digits could be used to dramatically increase the tracking capabilities of the system.

Once an individual producer is assigned a unique identification number, that producer begins the process of animal identification. The producer assigns numbers to his animals in the same way that a bank assigns checks. A producer does not have to identify which animal received which number, he is only responsible for ensuring that the numbers are assigned sequentially.

On receipt of an identification number, a sequence of tags is issued to that user. This sequence is recorded in the master database and the individual user need not maintain additional records, unless he desires to do so. As an example, a user might wish to test for pathogens in his live animals. By referencing the identification number prior to performing any test, individualized information becomes accessible and the user is able to capture this information into his personal database for later transfer to the master record.

As animals are sold, a record of the individual animal identification changes with title of ownership. Individual animal identification is required prior to transfer of ownership to ensure integrity of the identification record. This information is recorded and a complete record of the animal is available for reference.

If an identification tag is lost or damaged, a new tag is attached to the animal using the current owners' prefix and the next logical record of the second sequence. Reconciliation of the record is done in a similar fashion to standard inventory control systems.

BENEFITS TO GOVERNMENT

Animal ID/Trace Back

As hearings and debates continue, the need for a system of identifying animals is becoming more obvious. Consumer groups have begun to pressurise the food industry and government alike for an effective trace back program. Consumer groups have cited the need for a trace back system for tracking residue violations and as a means of ensuring food safety.

A system of animal identification and trace back will enable regulatory, private and consumer needs to be addressed. Origen's$^{(TM)}$ ability to identify animals provides a system of trace back that no other system can offer. No other system is capable of identifying sources of contamination for events similar to the *E. coli* incident in the US. For a trace back system to be timely and effective, the product must be identifiable from the consumer back to the point of origin. Furthermore, Origen$^{(TM)}$ is also capable of effecting a recall of tainted product should it become necessary. This recall process is available if users employ the full capabilities of the system.

Cooperation/Communication

It is becoming more apparent that in order to compete in a global market, greater degrees of cooperation between industry and government are necessary. Establishing uniform goals and systems is the first step. Standardization of information systems among various government agencies is the beginning of this process.

The correlation and comparison of data between agencies and countries promotes a new level of cooperation. Data from one department may easily be transferred and disseminated, cutting research and investigation costs. As a result, many of the complaints regarding duplicity of effort and conflict of interest can be avoided. For instance, in the US, the same information required by the Food Safety Inspection Service (FSIS) for food safety can be coordinated with the Animal Plant Health Inspection Service (APHIS) and other agencies or governments.

Pathogen Tracking

Origen$^{(TM)}$ also has the distinct advantage of being able to correlate data in order to trace back pathogenic contamination to its source. In the case of pathogens that originate from a single source or location, a recall of animals associated with that source can be achieved. This immediately identifies and halts further exposure of the public's safety to these contaminants. Since pathogen contamination may not be detected until a product reaches the consumer, the ability to trace products back from the consumer is imperative.

Hazard Analysis and Critical Control Points (HACCP)

With the aid of an animal identification system, HACCP is more complete. Individual tests may be done on samples or on entire groups. Animal identification aids in the ability to correlate common traits among food products that demonstrate a high incidence of contamination. Such an ability greatly enhances data collection for risk assessment.

Tests may be further standardized so that animals of similar origin or other commonality are tested at specific control points. This greatly increases the reliability of certain HACCP procedures for conditions that have a high probability of contamination. Such a case may also include animals that are imported into a country. Since many current HACCP programs tend to overlook live animal production, an identification system can become a first step for ranchers desiring to implement HACCP controls at this phase of the production process.

Slaughter plants using animal identification systems, whose HACCP procedures have proven effective in controlling contamination, can share information with other facilities experiencing similar problems. Animal identification further aids pathogen testing techniques which do not quickly yield results prior to a carcass leaving the slaughter plant. If a carcass or carcasses were to test positive, they could be recalled even after leaving a slaughter plant should that become necessary.

And finally, with the ability to recall tainted products, current HACCP programs will be able to demonstrate their effectiveness to consumers. A recall process will go a long way to assuring consumers that a situation similar to the *E. coli* incident can be handled in a timely and effective manner. Had an effective recall system been in place, consumer confidence would not have been lost.

Imports

As free trade and open markets continue to develop, a system of identification across international boundaries becomes necessary. Current differences in safety regulations and production methods among trading partners are sure to result in problems. Potential problems with animal diseases entering a country may result as a consequence of free trade. The sophistication of current food distribution systems further complicates the issue. Animals may be slaughtered and shipped all over the world. A problem arising in one area of a country would be very difficult to trace through the entire distribution chain. Animal identification will enable monitoring and trace back throughout the entire distribution process.

International agreements and concerns are addressed with this identification system. Animals coming into a country are identified and traced through a single system. International trade relations are improved as producers are able to assure production practices for goods they export.

Imported foods that make specific product claims are difficult to verify for authenticity and safety. Origen's$^{(TM)}$ identification system makes it possible to verify the source and claims of any food product entering a country (provided Origen$^{(TM)}$ was in use prior to export.)

Inspection

In the US, the inspection process of animals prior to slaughter, and the resulting meat products, are presently under fire from both the public and the media. Government agencies are responding to this criticism as rapidly as they can, considering the current restrictions on manpower and finances. The FSIS is also being hindered by powerful industry forces that are justifying present methods and shifting blame to others.

With Track II, FSIS has created a new approach to inspection by balancing concerns of both industry and consumers. This plan will eventually need a means of animal identification in order to improve procedures, assure food safety to consumers, and provide a means of trace back.

The use of an animal identification system further aids the coordination of USDA inspection with plant HACCP procedures, resulting in improved efficacy of the inspection process.

Standardization

Throughout the food industry products are numbered and labelled by many different producers for a number of reasons. The standardization of these identification systems simplifies procedures. No longer will it be necessary to use conflicting identification methods to determine product identification. Standardization based upon a single electronic identification number further increases the timeliness of the information retrieval. Tracking products through manually kept records is inefficient, costly, and time consuming.

Attempts to correlate and collect data from different organizations is difficult. The problem is further complicated as the number of producers involved with a single product increases. A standardized numbering system for food products will facilitate the identification of individual parties involved in the identification process.

Certification

As more and more products seek to differentiate themselves from the competition, the role of governments to regulate and certify unique labelling claims becomes difficult. However, the call for greater regulation with lower budgets presents certain problems. One way to increase product certification, while maintaining costs, is to automate the certification process. If all information relative to a product were captured in a computer database, labelling claims could easily be verified.

The Origen$^{(TM)}$ system is capable of verifying specific claims made by producers. Certain product specifications, such as are needed for organic products, could be adopted and the producers required to validate such claims during production in the Origen$^{(TM)}$ record.

Modernization

A US Secretary of Agriculture recently said, "We cannot continue to run a system based on 1933 standards in 1993." Current inspection systems were not created to take advantage of the current available technological advances. The time has come to update our present system to take advantage of all useful technology that is available.

An observation of other industries should demonstrate to the food industry that the future lies in information systems. Identification of food products will happen sooner or later.

BENEFITS TO INDUSTRY

Farmers/Producers

Most farmers and stockmen are proud of their products and are eager to know how their products compare with others. They are also interested in methods that can improve their products. An integrated system of food identification allows them to make informed conclusions about their products.

An animal identification system can provide stockmen with valuable data about the genetic performance of their breeds. This data can be scientifically compared to other breeds for desired performance characteristics. Producers will be repaid for their effort in the form of products that bring a premium price.

A farmer would also be able to differentiate his products, for example, based upon production methods. A problem, such as the inability to export beef due to BSE contamination, can be averted by a producer's ability to demonstrate, by record, that no animals in his herd have ever been implicated in a BSE outbreak.

Feed Yards

Congestion is greatest at feed yards and, as a result, disease is more rapidly transmitted. It is also the place where residues of unwanted microbial or other materials can contaminate the actual meat supply. In order to track these animals, individual or lot identification is imperative should a problem arise in the resulting food produced by these animals. Feed yard operators are eager to make use of any information that can improve efficiency. This information may be relative to genetics, rates of consumption, or tracking actual feedstuffs back to their origin. The same process used to track animal pathogens can also be used by a feed yard operator to monitor his production efficiency.

Slaughter Plants

The slaughter plant occupies a key position in the meat production industry. It is the point at which live animals are converted to food. Here, important discoveries are made in relation to animal diseases and pathogens. At no other point in the meat production process is there a greater danger of contamination from outside sources.

Inspection procedures can be greatly enhanced by correlating organoleptic inspection with HACCP. Since some animal diseases are not found until the animal is slaughtered, animal identification enables slaughter plants to identify and trace potential sources of pathogens, should any contamination be detected. In the event that a pathogen is discovered, animals from similar feedlots, or other common traits can be tested to determine correlation for a cause of the contamination.

Slaughter plants can also achieve increases in production efficiency through the ability to control purchasing based upon desired animal characteristics. Since animal identification is transferred from the live animal to the meat product, slaughter plants can change their purchasing practices from the purchasing of live animals to the purchasing of meat.

Fabrication Plant

Good fabrication and processing plants are constantly searching for ways to improve production methods. Good food manufacturing companies ask themselves daily if their products are safe, of the highest possible quality, and if they satisfy consumer demand.

Origen[TM] has the ability to control inventory, provide production information and assist quality control. Purchasing decisions are further benefited by yield to price ratio analysis based upon the actual meat produced from an animal. A number of other accounting and production control measures benefit as well. For example, quality control departments will benefit from improved HACCP procedures based upon individual animal identification. Accounting departments will benefit from greater inventory control measures that an individual identification system provides.

If implemented correctly, Origen(TM) also allows fabricators the ability to recall problem products. As consumer demands for safer foods increase, fabricators can assure their customers of greater product integrity.

Distributors

Distributors everywhere deal mostly in packaged products. Inventory control can be enhanced with an integrated product identification system. This is especially true when dealing with commodity products where the source of origin is difficult to determine. Since Origen(TM) takes advantage of electronic identification, distributors can use this system to improve inventory control of their products.

Retail/Food Service

At the retail or food service level, the greatest amount of responsibility is assumed in dealing with the ultimate consumer. These entities are at the front of the firing line should a problem arise. They are also the ones who are best able to educate and inform consumers. In the event of a food safety problem, these parties will be the first to be approached.

The ability to identify and trace back a food product before it becomes a significant public food safety concern is important. The US fast food chain, Jack in the Box, would probably consider a system of product identification very worthwhile considering the bad publicity and losses associated with their involvement in the *E. coli* contamination problem.

BENEFITS TO CONSUMERS

Consumers are concerned about food safety. The Origen(TM) identification system assures consumers that the production process of food products are traceable. This tracking system ensures that the food production process is being checked and monitored for safety.

In the event that a problem escapes production safety controls, the product can be quickly identified. Once a product is identified as a potential problem to public safety, a rapid recall of similar products can be done. The ability to recall products will ensure that other consumers are not affected by the same problem.

SUMMARY

In conclusion, an integrated animal identification system, such as Origen$^{(TM)}$, has numerous benefits to consumers, producers, and governments alike. By adopting such a system, valuable information can be maintained and utilized to improve food safety and production efficiency. No other system possesses the ability to monitor food products from "Farm to Table" with trace back and recall abilities.

The Origen$^{(TM)}$ system of identification aids in the trace back of animal diseases and pathogens. Sources of contamination can be traced and corrected before they become a serious public health hazard.

The ability to share and coordinate information among industry and government can be achieved with the Origen$^{(TM)}$ system. By referencing a single identification number that is used by many agencies, duplicity of effort is removed and greater efficiency achieved.

Certification of unique product claims is easier to verify with the aid of this identification system. Specific claims are recorded and maintained through to the end consumer. Questions or concerns regarding product claims or production methods can be verified against the Origen$^{(TM)}$ record.

Private industry benefits from the Origen$^{(TM)}$ system by the additional information and control that it gains for its products. Additional value is inherent with products using an identification system. Consumers and food safety groups who are demanding greater controls for food safety can be assured of a valid recall ability for tainted products.

In short, no other system provides the ability to do so many things and to solve so many of the current concerns involved in food safety. In terms of the benefits that such a program offers, isn't it time that we considered implementing a system that will give us an integrated and comprehensive system to insure consumers that the food they purchase is safe for their families.

CHAPTER 5

THE USE OF RAPID DNA BASED PATHOGEN DETECTION METHODS IN THE DEVELOPMENT OF HACCP SYSTEMS

TERRY SMITH, MAJELLA MAHER and GERALDINE MOLLOY

National Diagnostics Centre, BioResearch Ireland
University College Galway, Ireland

ABSTRACT

The development of DNA probes for the rapid identification of pathogenic bacteria in foods is discussed. The basis of these probes is that DNA sequences, unique to a specific organism, are used which give a high degree of specificity. A development of major significance is the polymerase chain reaction which facilitates the amplification of the target DNA. This allows a rapid increase in the concentration of the target DNA to high levels, which facilitate its detection. The selection of the DNA to be used in a probe is critical. The use of ribosomal DNA is recommended and its advantages outlined. This approach has been used to develop DNA probes for the detection of Listeria *species. The* Listeria *DNA technology has been further developed to produce an enzyme immunoassay kit. The kit is presently being refined to reduce assay time. A further development is the use of an automated fluorescent sequencer which has been shown to be potentially a rapid and sensitive detection system. Finally, the use of DNA probes in the food industry in the future are presented in the context of their ultimate use in HACCP systems.*

INTRODUCTION

The rapid detection and identification of microorganisms is of vital importance in many areas including clinical and veterinary medicine, the environment, and in industries such as food production and processing. Many methods are used to identify microorganisms which usually involve observations on growth. These include classical microbiological techniques, involving culture on selective media followed by biochemical analysis. More recently automated microbiological methods, which detect growth of organisms in selective media at a very early stage, have also been developed, i.e., Bactec and Lumac. Another approach is the analysis of the fatty acid profile of microorganisms which provide a type of fingerprint pattern that varies from species to species (Moss 1985). Immunological methods are also being applied to the detection and identification of microorganisms, using specific antibodies directed against pathogen-specific antigens (Morris *et al.* 1988). A novel detection and identification approach, which is increasing in use and can avoid the need for culture of the target organism, involves the direct detection of the genetic material (DNA) of the organism in a specimen.

Development of DNA Detection Assays

Recent major technological advances in molecular biology have opened up many exciting new avenues of opportunity for the application of DNA diagnostics to the detection and identification of microorganisms, particularly bacteria for which standard laboratory diagnosis is difficult, time consuming or impossible. These procedures have now been extended to fungi of agricultural, food and environmental importance. In the food sector, these approaches will provide a very useful rapid assay system for the detection of foodborne pathogenic microorganisms during all stages of food production, preparation and processing. These assays are being refined and improved in speed, sensitivity and user friendliness in order to allow their application in food production situations, such as in hazard analysis of critical control points (HACCP) operations.

The basis of DNA assays for the detection and identification of microorganisms is that each organism has regions of DNA sequences that are unique to a particular genus, species and even subspecies. These unique sequences, once identified, are used as "DNA probes" for the specific identification of that organism. In a DNA based assay the genetic material of a pathogen is detected and is, therefore, independent of the immunological status of the organism. These assays detect the DNA of a particular organism but do not necessarily detect living organisms, a point which may be of major significance in applying this technology in HACCP programmes.

The advantages of DNA probe assays include their absolute specificity for particular organisms and the speed of detection. With a major technological breakthrough, the polymerase chain reaction (Saiki *et al.* 1985), which involves the *in vitro* amplification of DNA, the amount of target DNA available to a DNA detection assay can be greatly increased from a single molecule to over one million times its concentration in a sample. Thus, even a very small number of organisms present in a clinical specimen (Brisson-Noel *et al.* 1991; Cormican *et al.* 1992) or food sample (Flowers *et al.* 1987; Rossen *et al.* 1992) can be detected using the DNA probe technology. The specificity of the assay can be incorporated into the PCR amplification reaction and/or by hybridisation of the PCR product to a DNA probe, which is specific to the microorganism being tested.

Various strategies have been used to obtain DNA probes which specifically identify given bacteria. These include:

(1) Sequences coding for antigenic determinants characteristic of the organism e.g., Mycobacterial antigen MPB64 (Yamaguchi *et al.* 1989).
(2) Virulence factors e.g., listeriolysin from *Listeria monocytogenes* (Leimeister-Wachter *et al.* 1990).
(3) Sequences isolated by differential hybridisation e.g., *Aeromonas salmonicida* (Hiney *et al.* 1992).
(4) Sequences from randomly cloned DNA fragments.
(5) Ribosomal RNA (rRNA) gene and related sequences e.g., 16S region rRNA probes (Barry *et al.* 1990; Barry *et al.* 1991; Glennon *et al.* 1994).

The ideal target for a DNA probe should be both highly abundant and specific to the organism of interest. Ribosomal RNA satisfies the criterion of abundance with up to 10^4 copies of rRNA per cell. Ribosomal RNA genes in bacteria are arranged in transcription units or operons, consisting of three genes, coding for 16S, 23S and 5S rRNAs (Brosius *et al.* 1981). All three genes are co-transcribed as a single precursor RNA. Ribosomal RNA operons are present in multiple copies, e.g., seven operons in *E. coli* (Kenerley *et al.* 1977; Kiss *et al.* 1977) and nine in *Clostridium* (Canard and Cole 1989). During active cell growth and division, very high levels of rRNA are present. It has been known for some time that rRNA genes are conserved among diverse bacterial genera. Extensive sequence data on rRNA genes has shown recently that rRNA genes consist of regions which are highly conserved and others which are variable, even between closely related species (Dams *et al.* 1988). As RNA has a structural role rather than that of a template for translation, probes based on it will be more sensitive than those obtained by methods 1-4 above, unless amplification strategies such as PCR are employed. The capacity to amplify DNA *in vitro* will extend some of the uses of DNA probes but questions, such as the ease with which they can be integrated into routine use without problems of DNA contamination, require resolution before they can be used without very demanding precautions.

A general method has been developed in this laboratory to generate DNA probes from the 16S rRNA gene of bacteria (Barry *et al.* 1990).This strategy, based on PCR technology, involves the use of sequences from constant regions within the 16S rRNA gene, as PCR primers, to amplify across variable regions. In many cases, however, very little variation is seen between closely related species and even genera in the variable regions within the 16S gene. In order to circumvent this problem, an alternative strategy has been developed, based on the intergenic (spacer) region between the 16S and 23S rRNA genes (Barry *et al.* 1991). It was reasoned that evolutionary pressure would be less on noncoding DNA than on coding DNA, resulting in increased sequence variation. Using highly conserved sequences from the 3' end of the 16S rRNA gene and the 5' end of the 23S gene as PCR primers, the intergenic or "spacer" region can be amplified and the DNA sequence determined. Comparison of the rRNA spacer sequence with those of closely related organisms and those previously lodged in DNA sequence databases indicates regions of identity and regions of unique sequence, which allows species-specific DNA probes to be designed. This strategy has been used to develop DNA probes specific for a number of pathogenic bacteria including *Salmonella* species (Molloy and Smith 1995), *Mycobacterium tuberculosis* (Glennon *et al.* 1994) and *M. bovis*. A DNA probe for *Clostridium perfringens* has also been developed using this approach (Barry *et al.* 1991). An extension of this, leading to improved sensitivity, is to design PCR primers specific for an organism of interest from the DNA sequence data generated. This strategy has been used successfully to detect *Clostridium perfringens* when it is a minor contaminant in a sample under investigation.

DNA Detection Assay for *Listeria*

As part of a programme in developing DNA based assays for the detection of food pathogens, this approach has been applied to determine the nucleotide sequence of the spacer region of *Listeria monocytogenes, L. innocua* and *L. welshimeri* (Maher *et al.* 1995). Comparison of the sequences using computer alignment programmes (Higgins and Sharp 1988; Higgins *et al.* 1992) allowed the design of a *Listeria* genus-specific PCR amplification assay and DNA probe as well as an *L. monocytogenes* specific PCR assay and DNA probe. The limit of detection of the PCR and DNA probe assay was determined to be one bacterial cell using radioactive DNA probe hybridisation.

Development of an Enzyme Immunoassay DNA Detection System

There are however, major limitations with PCR based DNA assays at present, as they require specialised equipment, highly trained personnel and usually involve radioactive detection methods. In order to be of use in a food production process, the assays must be rapid, robust, easy to perform, sensitive and specific. As a first step toward making these assays more user friendly and suitable for use within the food industry, a microtitre plate enzyme immunoassay (EIA) format has been developed for the detection of PCR products (Molloy *et al.* 1995). This PCR product detection EIA is a competitive disequilibrium assay based on the detection of the binding of biotin to streptavidin (Fig. 1). PCR reactions are carried out using oligonucleotide primers which have a biotin molecule attached at their 5' end for detection purposes and which does not affect the efficiency of the PCR amplification reaction. The detection assay involves separate incubation periods for the biotinylated PCR products and a biotin horseradish peroxidase conjugate. There is competition between the biotinylated DNA and the biotin-enzyme conjugate for streptavidin binding sites.

COMPETITIVE DNA DETECTION EIA

Wells coated with Streptavidin
Biotinylated PCR product added

Biotin enzyme conjugate added

Colorimetric enzyme substrate added

Higher OD = Less PCR product = Less Target

FIG. 1. SCHEMATIC REPRESENTATION OF THE ENZYME
IMMUNOASSAY FOR THE DETECTION OF PCR PRODUCTS
Streptavidin molecules are bound to the bottom of the microtitre plate wells
and biotinylated PCR products bind to the streptavidin. A biotin HRP enzyme
conjugate is then added which binds to free streptavidin sites and a
chromogenic enzyme substrate is converted to a coloured compound by
streptavidin-bound enzyme conjugate.

As the concentration of biotinylated PCR product increases, more streptavidin
sites on the microtitre plate become bound, with the result that the amount of
biotin-enzyme conjugate that can bind available streptavidin sites decreases and
a lower optical density (O.D.) reading is obtained. Thus, the greater the amount
of bacterial DNA in a sample, the greater the amount of PCR product produced
that can bind to the streptavidin and a lower colour signal is observed. A biotin
standard curve is set up for all assays with known amounts of pure biotin and

all readings obtained from biotinylated PCR samples are related to this. The O.D. value obtained for the highest concentration of conjugate standard is referred to as the Bo value and this is used to calculate the percentage binding of PCR products to the streptavidin sites, referred to as the B/Bo value.

The sensitivity for detection of *Listeria*, using this assay detection format, was established by PCR amplification of serial dilutions of DNA, followed by EIA detection of the PCR products (Fig. 2). The limit of detection of PCR product achieved, to date, is approximately 10pg of target DNA or 2000 cells, which is significantly less sensitive than radioactive detection methods. This sensitivity has been achieved consistently for a variety of DNA targets (Molloy *et al.* 1995). Having established the limit of detection of DNA using the EIA, the ability to detect *L. monocytogenes* in food samples was established using

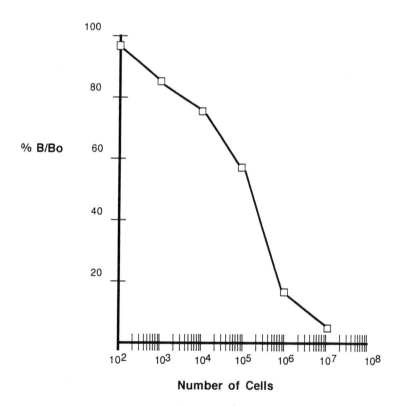

FIG. 2. SENSITIVITY CURVE FOR THE DETECTION OF *LISTERIA* CELLS
USING PCR, FOLLOWED BY EIA DETECTION
The % B/Bo indicates the amount of colour in a well relative to the biotin standard.
100% colour indicates the absence of DNA in a sample.

PCR amplification and EIA detection. Samples of milk (25.0 ml) spiked with 100 *L. monocytogenes* cells were transferred to 225 ml PBS, mixed well and 10 ml (equivalent to 4 cells) removed and inoculated into 225 ml of growth medium for culture. A 10 μl aliquot of the culture media was taken at various time points and subjected to PCR amplification, followed by EIA detection. *L. monocytogenes* organisms were detected after 20 h of culture in this assay, which in total took 27 h to perform. While this experiment established the feasibility of using a PCR and EIA assay for the detection of *L. monocytogenes* in food it was not optimised in any way. This assay protocol is now being refined in a number of ways to reduce the overall assay time. Improvements being introduced into the assay protocol are expected to reduce the culture time to 5-6 h and the overall assay time to 10-12 h.

Fluorescent PCR Product Detection on ALF

Work has also been carried out on an alternative strategy for the detection of PCR products which involves the detection of fluorescently labelled PCR products using an Automated Laser Fluorescent (ALF) sequencer (Pharmacia, Uppsala, Sweden). This automated electrophoresis and laser detection system, which was developed for DNA sequencing purposes (Voss *et al.* 1992) can also be used as a diagnostic PCR product detector. Laser beams in the ALF automatically detect fluorescein labelled DNA fragments which can be generated using fluorescein labelled nucleotides or PCR primers with a fluorescein molecule attached at the 5' end. Fragment size and analysis software is available to analyse the fluorescently labelled products of diagnostic PCR assays.

The sensitivity of the ALF detection system was evaluated using serial dilutions of DNA and cultured cells down to a single organism. Bacterial cell numbers were confirmed by plate counts. The results of this evaluation indicates that the detection of fluorescein labelled PCR products on the ALF is as sensitive as radioactive DNA probe detection and detects the presence of a single organism in a sample (Maher *et al.* 1995). The ALF fluorescence system for the detection of PCR amplified bacterial DNA is now being applied in clinical and food samples. Its advantages over the enzyme immunoassay format include sensitivity and speed, since PCR products can be loaded directly onto the electrophoresis apparatus without any purification or manipulation, as well as automatic data storage, retrieval and analysis software. In addition up to 40 samples can be analysed simultaneously which would be of advantage in a large sample throughput environment. Thus, the combination of PCR amplification and detection on the ALF could provide a rapid and sensitive pathogen detection system in a HACCP programme.

Future Developments

With these PCR based bacterial detection assays, the detection and identification of an infectious agent can be achieved within hours, which is of enormous assistance in speeding up management decisions in a food production process. This could provide a HACCP monitoring system for the food industry. In addition to allowing the detection of very little starting material in a sample with the use of *in vitro* DNA amplification systems, it is possible to simultaneously screen for multiple pathogens using a multiplex PCR assay. This involves the incorporation of specific PCR primers for each pathogen into the amplification process and their detection on a PCR fragment size basis. Some of the problems associated with the technologies used currently are being addressed for the requirements of the food industry. This involves investigating the use of isothermal DNA amplification systems which would simplify the *in vitro* amplification process enormously. While the ultimate goal is to have a kit format for use in HACCP which will give a result within minutes, the formats which will initially be utilised will require a culture enrichment step. This requirement will continue in the medium term, since the amplification systems are required to detect as little as a single organisms in 25 g (ml) of food. Until methods are devised that can separate the food components from nucleic acids and bacterial cells efficiently, some culture enrichment will be required.

Ultimately, DNA based assays which avoid the requirement for culture will come onstream with a bacterial purification step followed by *in vitro* amplification and detection. This type of assay will provide a very rapid detection assay for pathogens in a variety of areas including food production and processing. Once the industry accepts the technology of DNA based diagnostic assays and they become more user friendly, their incorporation into HACCP operations will automatically follow.

REFERENCES

BARRY, T., COLLERAN, G., GLENNON, M., DUNICAN, L.K. and GANNON, F. 1991. The 16S/23S ribosomal spacer region as a target for DNA probes to identify Eubacteria. PCR Methods and Applications, *1*, 51–56.

BARRY, T., POWELL, R. and GANNON, F. 1990. A general method to generate DNA probes for microorganisms. Biotechnology, *8*, 233–236.

BRISSON-NOEL, A., AZNAR, C., CHUREAU, C., NGUYEN, S., PIERRE, C., BARTOLI, M., BONETE, R., PIALOUX, G., GICQUEL, B. and GARRIQUE, G. 1991. Diagnosis of tuberculosis by DNA amplification in clinical practice evaluation. Lancet, *338*, 364–366.

BROSIUS, J., DULL, T.J., SLEETER, D. and NOLLER, H.F. 1981. Gene organisation and primary structure of a ribosomal RNA operon from *Escherichia coli*. J. Mol. Biol. *148*, 107–127.

CANARD, B. and COLE, S.T. 1989. Genome organisation of the anaerobic pathogen *Clostridium perfringens*. Proc. Nat. Acad. Sci. *86*, 6676–6680.

CORMICAN, M.G., BARRY, T., GANNON, F. and FLYNN, J. 1992. Use of polymerase chain reaction for early identification of *Mycobacterium tuberculosis* in positive cultures. J. Clin. Path. *45*, 601–604.

DAMS, E., HENDRIKS, L., VANDER PEER, Y., NEEFS, J.M., SMITS, G., VANDENBEMPT, I. and de WACHTER, R. 1988. Compilation of small ribosomal subunit RNA sequences. Nucl. Acids Res. *16* (suppl), 87–173.

FLOWERS, R.S., KLATT, M.J., MOZOLA, M.A., CURIALE, M.S., GABIS, D.A. and SILLIKER, J.H. 1987. DNA hybridisation assay for detection of *Salmonella* in foods: Collaborative study. J. Assoc. Off. Anal. Chem. *70*, 521–529.

GLENNON, M., SMITH, T., CORMICAN, M., NOONE, D., BARRY, T., MAHER, M., DAWSON, M., GILMARTIN, J.J. and GANNON, F. 1995. The ribosomal RNA intrergenic spacer region: a target for the PCR based diagnosis of tuberculosis. Tubercle, Submitted for Publication.

HIGGINS, D.G., BLEASBY, A.J. and Fuchs, R. 1992. Improved software for multiple sequence alignment. CABIOS, *8*, 189–191.

HIGGINS, D.G. and SHARP, P.M. 1988. A package for performing multiple sequence alignment on a microcomputer. Gene *73*, 237–244.

HINEY, M.P., DAWSON, M.T., SMITH, P.R., GANNON, F. and POWELL, R. 1992. DNA probe for *Aeromonas salmonicida*. Appl. Environ. Microbiol. *58*, 1035–1042.

KENERLEY, M.E., MORGAN, E.A., POST, L., LINDAHL, L. and NOMURA, M. 1977. Characterisation of hybrid plasmids carrying individual ribosomal ribonucleic acid transcription units of *Escherichia coli*. J. Bact. *132*, 931–949.

KISS, A., SAIN, B. and VENETIANER, P. 1977. The number of rRNA genes in *Escherichia coli*. Febs Letts, *79*, 77–79.

LEIMEISTER-WACHTER, M., HAFFNER, C., DOMANN, E., GOEBEL, W. and CHAKRABORTY, T. 1990. Identification of a gene that positively regulates expression of listeriolysin, the major virulence factor of *Listeria monocytogenes*. Proc. Nat. Acad. Sci., *87*, 8336–8340.

MAHER, M., DAWSON, M., GANNON, F. and SMITH, T. 1995. Unpublished data.

MAHER, M., DOWDALL, D., GLENNON, M., WALSHE, S., CORMICAN, M., WIESNER, P., GANNON, F. and SMITH, T. 1995. The sensitive detection of fluorescently labelled PCR products using an automated detection system. Mol. and Cellular Probes *9*, 265–276.

MOLLOY, G., MAHER, M. and SMITH, T. 1995. Unpublished data.

MOLLOY, G. and SMITH, T. 1995. Unpublished data.

MORRIS, B.A., CLIFFORD, M.N. and JACKMAN, R. 1988. In *Immunoassays for Veterinary and Food Analysis 1*. Elsevier Applied Science, London.

MOSS, C.M. 1985. Uses of gas-liquid chromatography and high pressure liquid chromatography in clinical microbiology. In *Manual of Clinical Microbiology*, Fourth edition (E.H. Lennette, A. Balows, W.J. Hausler Jr. and H.J. Shadowmy, eds.) pp. 1029–1036. American Society for Microbiology, Washington, DC.

ROSSEN, L., NOVSKOV, P., HOLMSTROM, K. and RASMUSSEN, O.F. 1992. Inhibition of PCR by components of food samples, microbial diagnostics assays and DNA extraction solutions. Int. J. Food Microbiol. *17*, 37–45.

SAIKI, R.K., GELFAND, D.H., STOFFEL, S., SCHARF, S.J., HIGUCHI, R., HORN, G.T., MULLIS, K.B. and ERLICH, H.A. 1988. Primer directed enzymatic amplification of DNA with a thermostable DNA polymerase. Science *239*, 487–491.

VOSS, H., WEIMAN, S., WIRKNER, U., SCHWAGER, C., ZIMMERMANN, J., STEGGEMANN, J., ERFLE, H., HEWITT, N.A., RUPP, T. and ANSORGE, W. 1992. Automated DNA sequencing system resolving 1,000 bases with fluorescein-15-dATP as internal label. In *Methods in Molecular and Cellular Biology 3*, 153–155.

YAMAGUCHI, R., MATSUO, K., YAMAZAKI, A., ABE, C., NAGAI, S., TERASAKA, K. and YAMADA, T. 1989. Cloning and characterisation of the gene for immunogenic protein MPB64 of *Mycobacterium bovis* BCG. Infect. Immun. *57*, 283–288.

CHAPTER 6

ESTABLISHMENT OF BASELINE DATA ON THE MICROBIOTA OF MEATS

ANN MARIE McNAMARA

Food Safety and Inspection Service, Microbiology Division
United States Department of Agriculture, Washington, DC 20250

ABSTRACT

The United States Department of Agriculture's Food Safety and Inspection Service has initiated the National Microbiological Baseline Data Collection Programs to determine the prevalence and levels of selected bacteria on meat and poultry slaughtered under federal inspection. This paper reviews the methodology and results of the first completed study on samples collected from steer and heifer carcasses. Potential applications for this type of testing are discussed.

INTRODUCTION

One of the major functions of the Microbiology Division of the United States Department of Agriculture's Food Safety and Inspection Service (FSIS) is to conduct microbiological surveys of meat and poultry products produced under Federal inspection. The Microbiology Division has extensive experience in collecting bacteriological data from beef carcasses during slaughter and dressing operations. The purpose of this paper is to compare and contrast the results of two microbiological data collection programs: the Nationwide Beef Microbiological Baseline Data Collection Program for Steers and Heifers and the Critical Control Point (CCP) Microbiological Verification Testing Pilot Program for Beef Slaughter Operations.

The Nationwide Beef Microbiological Baseline Data Collection Program for Steers and Heifers (Steer Heifer Program)

FSIS conducts specialized studies to provide microbiological information on raw products under its jurisdiction. The first of these studies was the Nationwide Beef Microbiological Baseline Data Collection Program for Steers and Heifers. The Steer and Heifer Program was designed to collect data for developing and maintaining a general microbiological description, or 'profile,' of fed cattle (steer and heifer) carcasses for selected bacteria of varying public health concern and for selected indicator organisms. This nationwide, nonregulatory program was unprecedented in its scope since data were collected on six pathogenic bacterial species and three groups of indicator organisms. When bacteria were found on carcass samples, the levels of bacteria present were enumerated.

FSIS chose steers and heifers as the target population for this study because they constitute 80% of beef animals slaughtered in the U.S. and are the immediate source of most retail cuts. All establishments that slaughter an average of approximately 40 or more fed cattle per week were included in the sample frame. There are approximately 100 establishment in this category. These establishments account for more than 99% of all steers and heifers slaughtered in federally inspected plants. From October 1992 through September 1993, tissue samples representing 2,089 steer or heifer carcasses were collected from establishments operating under Federal inspection. A sample size of about 2,100 carcasses ensured reasonable levels of precision for yearly estimates given the expected prevalence of the bacteria included in this study.

Samples were collected by FSIS inspectors-in-charge following detailed instructions and procedures. Samples consisted of subsamples taken from the rump, flank and brisket of a randomly selected carcass half that had cooled in the cooler for at least 12 hs. The rump, brisket, and flank were chosen for this program because these locations are most likely to become contaminated during the slaughter/dressing procedure. Each subsample consisted of a surface tissue section approximately 1 centimeter (0.5 in.) deep and comprising about 300 square centimeters (about 6 in. by 8 in.). Subsamples were separately bagged. The bags were placed in insulated shippers with gel packs capable of maintaining refrigeration temperatures and shipped to the designated laboratory via an overnight delivery service for processing.

The tissue samples were analyzed for the presence of those bacteria most often associated with human illness as determined by foodborne illness reports, other pathogens of interest because of the severity of human illness they produce, and certain bacteria, or groups of bacteria, thought to be indicators of general hygiene or process control (NRC 1985). From these carcass samples, over 18,000 laboratory analyses were conducted.

Viable aerobic bacteria (Aerobic Plate Count @ 35C) were recovered from the surface of 98.9% of the 2,089 carcasses tested in this program (Fig. 1, FSIS 1994). Coliform bacteria were recovered from 16.3% of 2,089 carcasses and *E. coli* (Biotype I) was recovered from 8.2% of 2,089 carcasses. *S. aureus* was recovered from 4.2% of 2,089 carcasses, Salmonella was recovered from 1.0% of 2,089 carcasses, *C. perfringens* was recovered from 2.6% of 2,079 carcasses, *C. jejuni/coli* was recovered from 4.0% of 2,064 carcasses, *L. monocytogenes* was recovered from 4.1% of 2,089 carcasses, and *E. coli* O157:H7 was recovered from 0.2% of 2,081 carcasses.

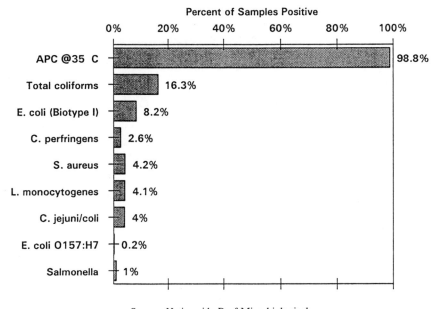

Percent of Samples Positive

- APC @35 C — 98.8%
- Total coliforms — 16.3%
- E. coli (Biotype I) — 8.2%
- C. perfringens — 2.6%
- S. aureus — 4.2%
- L. monocytogenes — 4.1%
- C. jejuni/coli — 4%
- E. coli O157:H7 — 0.2%
- Salmonella — 1%

Source: Nationwide Beef Microbiological
Baseline Data Collection Program: Steers
& Heifers (October 1992-September 1993)

FIG. 1. PREVALENCE OF SELECTED MICROORGANISMS ON RAW BEEF
CARCASS SURFACE SAMPLES

On carcasses that tested positive, the geometric mean of the Aerobic Plate Count @ 35C was 474.7 cfu/cm^2, the geometric mean of coliforms was 35.3 cfu/cm^2 and the geometric mean of *E. coli* (Biotype I) was 35.3 cfu/cm^2 (Fig. 2). When positive for a specific pathogen, the geometric mean on carcasses was: 24.3 *S. aureus* cfu/cm^2; 0.1 *Salmonella* MPN/cm^2; 45.1 *C. perfringens* cfu/cm^2; 0.1 *C. jejuni/coli* MPN/cm^2; 0.2 *L. monocytogenes* MPN/cm^2; and 0.6 *E. coli* O157:H7 MPN/cm^2.

The highest level detected (Table 1) for each of the various pathogens was: <100,000 cfu/cm^2 for *S. aureus*; <1.0 MPN/cm^2 for *Salmonella*; <100,000 cfu/cm^2 for *C. perfringens*; <1.0 MPN/cm^2 for *C. jejuni/coli*; and <1.0 MPN/cm^2 for *E. coli* O157:H7. Four of the 82 positive samples enumerated for *L. monocytogenes* reached the upper detection limit for the method of >11 *L. monocytogenes* MPN/cm^2. No further enumeration was done on these samples; the most probable next number, 24, was used in calculations.

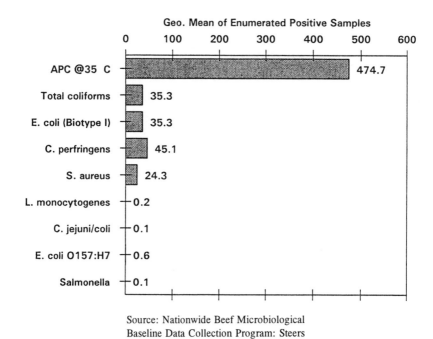

Source: Nationwide Beef Microbiological
Baseline Data Collection Program: Steers
& Heifers (October 1992-September 1993)

FIG. 2. MEAN LEVEL OF SELECTED MICROORGANISMS PER SQUARE CENTIMETER
ON RAW BEEF CARCASS SURFACE SAMPLES

TABLE 1.
HIGHEST LEVEL OF PATHOGENIC BACTERIA ENUMERATED ON RAW
BEEF CARCASS SURFACE SAMPLES

Microorganism Centimeter	Number/cm^2
Clostridium perfringens	< 100,000
Staphylococcus aureus	< 100,000
Listeria monocytogenes	< 100
Campylobacter jejuni/coli	< 1
Escherichia coli O157:H7	< 1
Salmonella species	< 1

Source: Nationwide Beef Microbiological Baseline Data Collection Program: Steers & Heifers
(October 1992-September 1993)

    Pathogens were not recovered from 1,785 (85.4%) of the 2,089 carcasses
tested (Fig. 3). 278 carcasses contained only one pathogenic bacterial species,
whereas 23 carcasses contained two species and three carcasses contained a total
of three species. No carcasses tested contained more than three pathogenic
species.

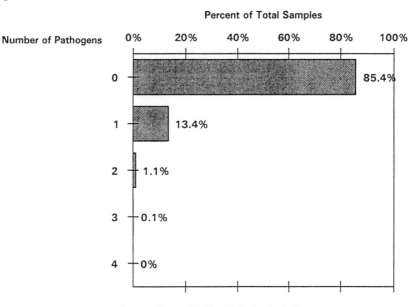

Source: Nationwide Beef Microbiological
Baseline Data Collection Program: Steers
& Heifers (October 1992-September 1993)

FIG. 3. PERCENT OF RAW BEEF SAMPLES CONTAINING ONE OR MORE SPECIES
OF IDENTIFIED PATHOGENIC BACTERIA

    Several important lessons can be learned from summarizing these data on
the presence and levels of pathogenic bacteria and indicator organisms on steer
and heifer carcasses. First, to detect the presence of pathogens on raw meats,
100% sampling of all pertinent pathogens on all samples collected must be
performed. This is due to the fact that when pathogens were recovered, they
were found infrequently and most often singularly in a product. This results in
very extensive, time-consuming (up to 7 days), and often costly testing
programs. Second, since the numbers of pathogens when recovered were below
the detection limits of current ELISA and DNA probe screening tests (104
organisms or greater), enrichment and selective broths are needed to detect
pathogens on raw products. This makes on-line testing for pathogens impossible
with current technology. Third, pathogenic bacteria were recovered infrequently

and generally in low numbers, therefore, testing raw meats for pathogens has little value in determining trends in process control programs. Fourth, indicator organisms such as the APC @ 35C, total coliforms, or *E. coli* (Biotype I) make better predictors for process control for they can be found in levels high enough to enumerate. The data from this study is being further evaluated to determine the relationship, if any, of the indicator organisms to the recovery of pathogens.

Critical Control Point Microbiological Verification Pilot Testing Program of Beef Slaughter Operations

The use of nondestructive sampling of raw beef carcasses is being evaluated in a new FSIS pilot testing program called the Critical Control Point Microbiological Verification Pilot Testing Program of Beef Slaughter Operations (CCP Verification Program). The objective of this study is to train FSIS inspectors in the collection of non-destructive microbiological samples and to evaluate the feasibility of collecting microbiological samples during inspection operations.

FSIS inspectors in five beef slaughter establishments are being provided with a sample collection kit consisting of a phosphate buffer solution (for rehydrating the sponge), a hydratable sponge, a transport media (Letheen broth), marking pens, sterile gloves, and sterile plastic bags. The rehydrated sponge is used to sample an area of the carcass that is approximately 300 cm² (3" x 18") located adjacent to the midline cut and including the brisket area. This area was chosen as it is an area which is most likely to become contaminated through deviations in dressing techniques. The size of the area is large to increase the likelihood of finding contamination. Carcass sampling involves repetitive sampling of the same area of the same carcass as the carcass progresses through processing. Samples are collected from the same carcass immediately after skinning, evisceration, final rinse, and chilling (National Advisory Committee 1994). Sample sponges are placed in sterile bags, the transport media added, the bags are labeled with pertinent sample information, and shipped refrigerated via overnight delivery service to FSIS laboratories for analysis. Each sample is being evaluated for total coliforms, *E. coli* (Biotype I), and aerobic plate count @ 35C by standard Association of Official Analytical Chemists (AOAC) procedures.

Although this study is not completed, there are several preliminary observations that can be made. First, this type of nondestructive sampling can show microbial trends in APC @ 35C, total coliforms and *E. coli* (Biotype I) numbers as carcasses move through a slaughter plant. Second, sponge sampling is simple for the inspector to perform, and, unlike destructive sampling, does not result in product loss. Third, this type of sampling is amenable to collecting large numbers of samples. However, early time estimates show that the collection of five independent samples and packaging for shipping is requiring approximately 1.5 h. Fourth, since these samples are shipped to FSIS laboratories for analysis, these sampling programs are applicable only to retrospective process control verification programs not to real-time, on-line process control monitoring.

Application to HACCP Programs

The results and observations of these two FSIS studies confirm some of the basic principles of microbiological testing in HACCP programs. Although a nationwide survey of raw beef carcasses was used to provide information relative to the occurrence of pathogens, it is important to understand that HACCP systems do not use microbiological data to analyze products for pathogens. Instead, HACCP systems are designed to provide information on the degree of process control exercised by a food manufacturer to reduce or eliminate microbiological hazards. The low frequency of occurrence and low level of pathogens on raw beef carcasses makes testing exclusively for pathogens an inappropriate means of evaluating process control. Indicator organisms are better predictors of process control if they are present on the product at levels high enough to enumerate.

In designing HACCP systems, the monitoring of CCPs by microbiological testing is retrospective due to the length of time required to obtain analytical results (National Advisory Committee, 1992). This fact makes microbiological testing applicable only to HACCP verification testing programs until new on-line, in-plant continuous microbiological monitoring technologies are developed. Until these new technologies emerge, current monitoring of CCPs is best achieved through visual observations and the use of physical and chemical tests.

REFERENCES

Food Safety and Inspection Service, Microbiology Division. 1994. Nationwide Beef Microbiological Data Collection Program: Steers and Heifers. USDA. Washington, DC.

National Advisory Committee on Microbiological Criteria for Foods. 1992. Hazard Analysis and Critical Control Point System. Int. J. Food Microbiol. 16, 1–23.

National Advisory Committee on Microbiological Criteria for Foods. 1994. Generic HACCP for Raw Beef. USDA. Washington, DC.

Subcommittee on Microbiological Criteria, Committee on Food Protection, Food and Nutrition Board, National Research Council. 1985. An Evaluation of the Role of Microbiological Criteria for Foods and Food Ingredients. National Academy Press. Washington, DC.

CHAPTER 7

THE ROLE OF INDICATOR SYSTEMS IN HACCP OPERATIONS

JAMES J. SHERIDAN

Meat Technology Department
Teagasc, The National Food Centre
Dunsinea, Castleknock, Dublin 15, Ireland

ABSTRACT

Many chemical and microbial indicators can be used to determine the safety and shelf-life of meats. This paper considers their potential application as elements in an effective HACCP system. A range of chemical indicators, including lactic and acetic acids and citrulline are examined. Many of these are of limited value, although citrulline offers some potential. Microbially based indicators and systems including, a contamination index and the use of temperature function integration are discussed. The correlation to the growth of pathogens is outlined, with particular reference to the modelling of E. coli. The potential value of psychrotrophic and psychrophilic counts as process hygiene indicators in the prediction of meat and keeping quality is elucidated. The application of these systems in HACCP is presented, with particular emphasis on their value in monitoring and verifying critical control points. The value of developing more effective methods using rapid techniques in this area is highlighted.

INTRODUCTION

The clinical and commercial consequences of food poisoning, as demonstrated by the fatalities caused by *E. coli* O157:H7 in the USA and by a number of similar cases involving *Salmonella* (Ryan *et al.* 1987; Roberts *et al.* 1989; Dorn 1993) present increasing problems to food processors. The occurrence of these outbreaks have led many in the food industry to conclude that traditional methods, such as inspection and end product testing, cannot provide adequate food safety. Attention has thus focused on the development and application of comprehensive quality assurance programs such as Hazard Analysis and Critical

Control Points (HACCP) to control food contamination during production (ICMSF 1988).

HACCP is a system which breaks down food handling/processing into its component steps, identifies the potential hazards associated with each step, and establishes specific actions to be taken at each stage (Critical Control Points) to protect food safety. It is the most important system of food safety management (Mortimore and Wallace 1994) and is currently being introduced into a large number of food processing operations, including meat and meat processing (Clarke 1989; Hathaway and Bullians 1992). An effective HACCP system needs to apply effective indicators in the selection, monitoring and verification of critical control points.

Both biological and chemical indicators are available for application within HACCP systems. In the first case, indicator organisms can be used. They are a group or species of organisms whose presence suggests hazardous conditions may have been encountered. Specific indicators may be used to show excessively contaminated raw materials, unsanitary manufacturing practices, unsuitable time/temperature storage conditions or process failure (ICMSF 1982; Mossel and Van Netten 1991). The concentration of chemicals produced by microbial activity may also give an indirect indication of the nature and numbers of organisms present.

CHEMICAL INDICATORS

According to Fields and Richmond (1968) a chemical indicator should: be present at low levels or be absent in uncontaminated food, increase in concentration as microbial growth increases, make it possible to differentiate low quality raw materials from poor processing conditions and be produced by the dominant microflora. Ideally, metabolic indicators should also be capable of being used predictively and their concentration should relate to the numbers of food-borne pathogens (Jay 1986). This would eliminate the possibility that food containing dangerous levels of pathogens, but a relatively low overall bacterial load, would be passed for human consumption. While this criterion is desirable, it is not practicable; no one indicator can cover the growth of all pathogens. In relation to raw meats a number of metabolites have been considered as indicators of process failure, including lactic acid, amines, acetic acid, amino acids, ammonia, and extract release volume (ERV). These are discussed individually below.

Lactic Acid

Lactic acid may be present in fresh meat in two isomeric forms, i.e., L (+) and D (-). L (+) lactic acid occurs naturally in high concentrations (1620–3800

$\mu g/g$) in fresh meats as an end product of glycolysis (Kakouri and Nychas 1994; Borch and Agerhem 1992; Nychas and Arkoudelos 1990). Its presence is therefore unrelated to microbial activity. The D (-) form of lactic acid is usually absent or present at very low levels in fresh meats (de Pablo *et al.* 1989; Borch and Agerhem 1992) and is produced in packaged meats solely by the activity of the lactic acid bacteria (Shaw and Harding 1984).

Increases in the lactic microbiota are reflected by increases in D (-) lactic acid concentration, as shown in Fig. 1 for vacuum and modified atmosphere packaged pork. The data show a good relationship between the concentration of D (-) lactic acid and the lactic acid bacteria present. Differences in the goodness of fit between the vacuum packaged product and the oxygen enriched atmospheres are reflected by differences in the regression coefficients. The superior R^2 value for the vacuum packs is caused by the stronger growth of the lactic acid bacteria, compared to that in the modified atmospheres where organisms such as *Brochothrix thermosphacta* also occur in high numbers (Taylor *et al.* 1990). D (-) lactate concentrations greater than $\log_{10} 3.0 \mu g/g$ are indicative of spoiled product (de Pablo *et al.* 1989). The production of D (-) lactic acid is also related to the growth of the lactic acid bacteria on beef (Borch and Agerhem 1992). This metabolite has been suggested as a suitable indicator of quality for many meat products (Sinell and Luke 1978; Schneider *et al.* 1983). The availability of a simple and enzymatic kit for quantifying D (-) lactic acid facilitates its use as an indicator (Anon. 1991).

Biogenic Amines

Biogenic amines may be formed by microbial activity on meat. The diamines putrescine and cadaverine are potential indicators of spoilage (Slemr 1981; Edwards *et al.* 1985) and may be indicators of aerobic spoilage in beef and pork (Sayem El Daher *et al.* 1984; Slemr 1981; Wortberg and Woller, 1982). In their investigations on aerobically stored beef, pork and lamb, Edwards *et al.* (1983) showed that the levels of putrescine (0.4–2.3 $\mu g/g$) and cadaverine (0.1–1.3 $\mu g/g$) were similar to those reported by other workers. However, these substances could not be applied as spoilage indicators because diamine concentrations were not significant until the bacterial counts had reached $4.2 \times 10^7/cm^2$, i.e., the meat was spoiled. This view has recently been substantiated by experiments on chicken skin aerobically stored at 4°C. Although absent initially, both amines were only present at very low levels (1.2 $\mu g/g$) when spoilage occurred after 6 days (Schmitt and Schmidt-Lorenz 1992a).

Putrescine and cadaverine may be useful indicators in packaged pork, chicken and beef (Edwards *et al.* 1985; Ordóñez *et al.* 1991; Schmitt and Schmidt-Lorenz 1992a). The levels of putrescine and cadaverine on chicken skin and pork loins stored in CO_2 enriched atmospheres are shown in Table 1. The

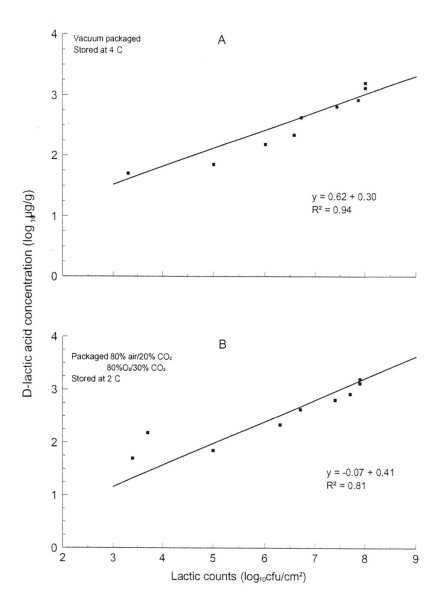

FIG. 1. RELATIONSHIP BETWEEN THE LACTIC COUNTS AND THE CONCENTRATION
OF LACTIC ACID IN PORK AND THE EFFECT OF PACKAGING AND
STORAGE TEMPERATURE
Data: A - Ordóñez et al. (1991). B - DePablo et al. (1989).

TABLE 1.
CONCENTRATIONS OF PUTRESCINE AND CADAVERINE (μg/g) IN PORK
AND CHICKEN SKIN AND THE RELATIONSHIP TO ENTEROBACTERIACEAE
COUNTS ($\log_{10}$ cfu/cm^2)

Storage time (days)	Pork[1]			Chicken skin[2]		
	Ent	Put	Cad	Ent	Put	Cad
0	0.2	0.2	0.1	2.88	0.0	0.0
2	1.0	0.4	0.4	2.88		
4	1.9	0.4	2.5	2.88	2.2	12.6
6				4.00		
7	2.2	0.2	0.6			
8				5.22	17.7	98.7 ←
10				6.00		
11	3.7	—	5.2			
→ 12				6.80	26.5	84.6
15	4.2	1.9	8.7			
16				6.80	48.9	161.4
18	5.8	10.6	21.1			
20				7.80	55.7	201.4
21	5.4	12.3	36.4			

[1]   Pork stored in 20% CO_2/80% air and 20% CO_2/80% O_2 at 2°C (Data: Ordóñez et al. 1991).
[2]   Chicken carcasses stored in impermeable packaging (approximate final gas composition 8% CO_2/12% O_2/80% N_2) at 4°C (Data: Schmitt and Schmidt-Lorenz 1992a).
Arrows indicate incipient spoilage time.
Ent = Enterobacteriaceae; Put = putrescine; Cad = cadaverine

data show that there is a higher concentration of cadaverine than putrescine, for both pork and chicken skin. The amounts of cadaverine in chicken skin are much greater, when the bacterial counts are about equal (pork - $\log_{10}$ 5.8 cfu/cm^2 - 21.1 μg/g; chicken - $\log_{10}$ 5.22 cfu/cm^2 - 98.7 μg/g). This may have been a reflection of differences in the microflora, storage temperature or in the gas atmospheres or combinations of these. These amines are linearly related to the growth of the Enterobacteriaceae ($R^2 = 0.89$) in pork (Ordóñez et al. 1991). While the biogenic amines may have potential, their widespread use seems unlikely. Their detection and quantification involves the use of specialist chemical techniques, which may preclude their detection and/or estimation in most routine circumstances. (Edwards et al. 1983; Schmitt and Schmidt-Lorenz 1992a).

Acetic Acid

Acetic acid may have potential as an indicator of microbial spoilage in packaged meats (Ordóñez *et al.* 1991). Acetic acid is produced by many organisms, especially the lactic acid bacteria or *Brochothrix thermosphacta* (Borch and Agerhem 1992; Dainty and Hibbard 1983). A good relationship between the production of acetic acid and bacterial numbers has been reported for the growth of the Enterobacteriaceae on pork (Fig. 2) (Ordóñez *et al.* 1991). A relationship between acetic acid levels and bacteria on the meat surface have been shown for beef and chicken (Borch and Agerhem 1992; Kakouri and Nychas 1994). In addition, the rate of acetic acid formation depends on the species of meat and whether it is dark or light within the same species (Kakouri and Nychas 1994).

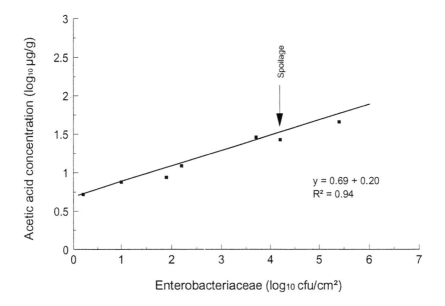

FIG. 2. RELATIONSHIP BETWEEN THE ENTEROBACTERIACEAE COUNTS AND ACETIC ACID CONCENTRATION IN PORK PACKAGED IN CO_2 ENRICHED ATMOSPHERES AND STORED AT $2\,^{\circ}C$
Data: Ordóñez *et al.* (1991).

In assessing the usefulness of acetic acid as an indicator, some further work is needed to resolve such issues as differences in the base levels in different foods. For example, acetic acid levels of 42.04–72.06, $\mu g/g$, 5.0 $\mu g/g$, 38.0 $\mu g/g$ and 50 $\mu g/g$ have been reported for beef, pork, chicken breast and chicken thigh, respectively (Borch and Agerhem 1992; Ordóñez et al. 1991; Kakouri and Nychas 1994). In beef the initial concentration range is so wide that the use of acetic acid as an indicator may be of limited value. While the detection of short chain fatty acids are commonplace, the extraction and detection procedures requires expert knowledge. The presence of a laboratory setup for routine analysis may also be needed, since once-off determinations would be very costly.

Amino Acids

Free amino acids occur naturally in fresh meat and the background levels vary depending on the meat type and age (Adamcic and Clark 1970). In fresh chicken skin, free amino acids represent about 1% of all available amino acids, while the levels in beef are less than half this amount (Jay and Kontou 1967; Lawrie 1985; Schmitt and Schmidt-Lorenz 1992b). The use of amino acids as an indicator of aerobic spoilage has been suggested by a number of authors (Gardner and Stewart 1966; Schmitt and Schmidt-Lorenz 1992b). The latter authors have shown that on chicken skin stored under aerobic conditions, but not in a CO_2 enriched atmosphere, there is a relationship between the concentration of free amino acids and the total counts.

There have been studies of citrulline as a specific indicator of meat spoilage including applications in both chicken skin packaged in air and in a CO_2 enriched atmosphere (Table 2) where the citrulline concentration increased in relation to the total counts (Schmitt and Schmidt-Lorenz 1992b). Citrulline is an ideal indicator, since it is not initially present in chicken skin. In both meat and skin the presence of citrulline arises from the activity of bacterial growth only. The presence of citrulline can be closely associated with the Pseudomonads. This amino acid arises from the breakdown of arginine to give putrescine and citrulline. Since citrulline production requires the growth of the Pseudomonads, the use of this substance as an indicator is limited to aerobic conditions. This is shown in Table 2 where the growth of the Pseudomonads would have been limited in the CO_2 enriched atmospheres. In consequence, from 8 days onward citrulline production ceases as Pseudomonads decline (Studer et al. 1988).

The use of the amino acids as indicators of aerobic spoilage is limited. They may have potential in particular cases as outlined above for citrulline. Usually they are unsatisfactory because their levels may increase or decrease with time (Jay and Kontou 1967; Saffle et al. 1961). Furthermore in many instances the levels of free amino acids may not reach significantly high levels until the meat is spoiled (Schmitt and Schmidt-Lorenz 1992b).

TABLE 2.

THE USE OF CHANGES IN CITRULLINE CONCENTRATION (μg/g) AS AN INDICATOR
OF MICROBIAL SPOILAGE (TOTAL VIABLE COUNT, TVC, $\log_2$/cfu/cm^2) IN CHICKEN
CARCASSES STORED IN AIR AND A MODIFIED ATMOSPHERE AT 4°C

Storage Time (days)	Air		Modified Atmosphere[1]	
	citrulline	TVC	citrulline	TVC
0	0.0	3.0	0.0	3.5
4	3.0	4.8	12.0	5.0
8	85.0	7.5	111.0	7.0
12	193.0	9.0	100.0	7.8
16	205.0	9.5	115.0	7.9

[1] Carcasses in polyethylene foil packs 8.0% CO_2
Arrow indicates incipient spoilage time
Data: Schmitt and Schmidt-Lorenz (1992b)

Ammonia

Ammonia may be an indicator of aerobic spoilage of meats (Lea *et al.* 1969; Rogers and McCleskey 1961). The Pseudomonads produce ammonia aerobically from fresh chicken and beef (Lea *et al.* 1969; Gill and Newton 1980; Nychas and Arkoudelos 1990), but its potential as a spoilage indicator is doubtful. As shown in Fig. 3 for chicken carcasses stored aerobically in desiccators, levels of ammonia only begin to increase significantly between 7 and 8 days, by which time spoilage has occurred (Schmitt and Schmidt-Lorenz 1992a). This has been observed previously by Lea *et al.* (1969) in chickens stored aerobically at 1°C. Schmitt and Schmidt-Lorenz (1992a) determined that the initial concentrations of ammonia varied from 60 to 130 μg/g. The large variation in the initial background levels of ammonia is a further drawback to its use as an indicator (Pearson 1968; Gill and Newton 1980).

Extract Release Volume

Extract release volume (ERV), although not strictly a chemical technique, is a rapid method for determining the microbial status of aerobically stored meat (Jay 1966; Borton *et al.* 1968). The relationship between ERV and microbial spoilage is shown in Fig. 4. This shows a linear relationship between these variables and that spoilage occurs at an ERV of about 24.0 ml, when the bacterial counts are at about $\log_{10}$ 8.0 cfu/g. This value of 24.0 is in fact the mean recommended number to indicate incipient spoilage (Jay 1966).

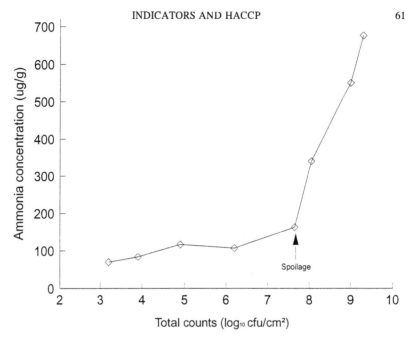

FIG. 3. AMMONIA CONCENTRATION IN CHICKEN CARCASSES STORED
AEROBICALLY AT 4°C
Data: Schmitt and Schmidt-Lorenz (1992a).

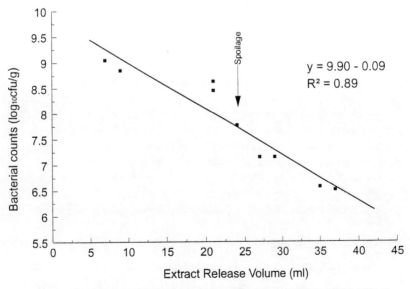

FIG. 4. RELATIONSHIP BETWEEN EXTRACT RELEASE VOLUME (ERV) AND
BACTERIAL COUNTS ON BEEF MINCE STORED AEROBICALLY AT 6°C
Data: Jay (1964).

ERV is very simple test to perform and can be carried out in only 15 min (Jay and Kontou 1964). The test is subject to errors due to variations between samples in the amount of fat present (Jay 1966). It is also limited by the fact that the mechanism of action is based on an interaction between the meat proteins and the bacteria present, which affects the ability of the meat to bind water. This reaction is not detectable at low bacterial levels and does not give meaningful results until the bacteria reach values of about $\log_{10}$ 6.0 cfu/g.

APPLICATION OF CHEMICAL INDICATORS IN HACCP OPERATIONS

Microbial metabolites might serve as indicators of spoilage in packaged meats where growth of certain organisms relates to the presence and/or concentration of a specific chemical substance. It is difficult to envisage these having anything other than a monitoring role. They could also be used to verify that an operation, such as storage, is ineffective in preventing spoilage and that the processes for control are not operating in a satisfactory way. For packaged meats, the diamine content has been used an indicator to determine the freshness of the product after storage (Edwards *et al.* 1985). For aerobically packaged meats, the tests are generally unsuitable indicators in HACCP systems because reliably detectable levels of the metabolites (free amino acid, ammonia, ERV) are not present until high bacterial counts have developed. Finally, many of the chemical indicators are precluded on the grounds of cost and complexity of operation. Indicators that are costly are of little value and will not be used in HACCP operations (Buchanan 1991).

MICROBIAL INDICATORS

Many methods can assess product quality based on specific characteristics of the resident microbiota. These are generally concerned with monitoring or verification of microbial numbers after different stages of processing and in some cases can be applied in establishing, monitoring and verifying critical control points for use in HACCP. Those of particular relevance to this paper include: process integrity indicators, temperature function integration, contamination index, and rapid psychrotrophic indicator systems.

Process Integrity Indicators

Process integrity indicators can verify that critical process parameters are within specification. According to Buchanan *et al.* (1991) process integrity is the integrated systems controls that are used to prevent pathogen growth. The system uses microbial indicators to assess the effectiveness of meat refrigeration.

This procedure can be used to detect temperature abused products. The method exploits the fact that temperature abuse of foods changes the ratios of mesophiles and psychrophiles present in the microbiota. These changes can be subsequently detected because of differences in the temperature optima of these two groups (Collins 1967).

When samples of ground beef were stored at 19°C (abuse) and 5°C (refrigeration) temperatures, plated out on brain heart infusion agar and incubated at 28 or 42°C, the counts allowed differentiation between materials stored at 19 and 5°C (Fig. 5A). This test also identified temperature abuse in a range of products, including chicken, pork and shrimp (Buchanan *et al.* 1992). In a further series of experiments, the higher incubation temperature for the plates was changed from 42 to 45°C (Buchanan and Bagi 1994). The data in Fig. 5B show differentiation of abused (19°C), refrigerated product (5°C) and also product that had been subject to temperature abuse as a result of fluctuating temperatures.

The psychrotrophic fraction (PF) developed by these authors shows the relationship between the proportions of psychrophiles and mesophiles in a product, and is expressed as:

$$PF = \frac{\text{Total count at } 28°C - \text{count at } 45°C}{\text{Total count at } 28°C}$$

A high PF value (0.35) indicates a higher proportion of psychrotrophic bacteria in the sample, while temperature abused products have lower PF values (0.15). The use of PF is shown in Fig. 5C. In relation to HACCP this type of process integrity indicator can be used as a verification technique. It could also be used as an alternative means of verifying raw ingredients specifications (Buchanan and Bagi 1994).

Temperature Function Integration

Temperature function integration (TFI) provides a means of predicting the microbiological status of meat from a record of its storage temperature. In this process the time intervals that the meat spends in each of a number of discrete temperature ranges is modelled. Data on the maximum number of bacterial generations which could have occurred on it during each of these intervals are correlated with this temperature information and summed to predict the total increase in bacterial numbers.

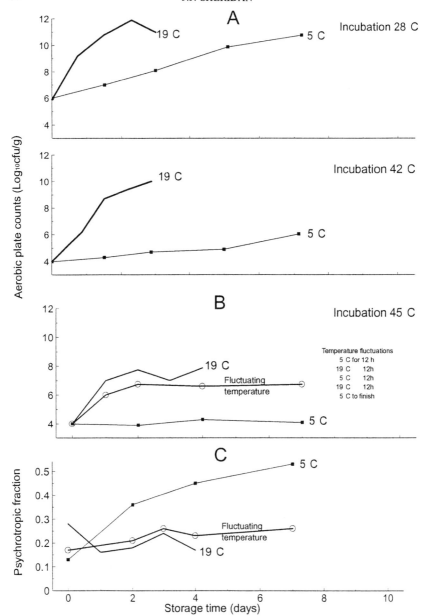

FIG. 5. A: THE EFFECT OF INCUBATING PLATES AT 28 AND 42°C ON TOTAL
COUNTS FROM GROUND BEEF STORED AT 19 (ABUSE) OR 5°C (REFRIGERATED).
B: THE EFFECT OF CHANGING THE INCUBATION TEMPERATURE TO 45°C AND
C: THE USE OF A PSYCHOTROPIC FRACTION AS AN INDICATOR OF
TEMPERATURE ABUSE
Data: Buchanan *et al.* (1992); Buchanan and Bagi (1994).

This approach has been used to predict the probable changes during carcass cooling, using the prediction of *E. coli* growth as an indicator for enteric pathogens on meat surfaces (Gill *et al.* 1991). *E. coli* is a suitable indicator as it is a frequent surface contaminant of meat and has growth patterns similar to pathogens, such as *Salmonella* (Mackey and Kerridge 1988).

For beef carcasses, the temperature profile is determined by attaching thermocouples to different sites on the carcass, such as the aitch bone cavity. The cooling data are down loaded to a computer which divides the derived cooling curve into a number of temperature ranges and resident times. Information on the growth rates of *E. coli* at each of these ranges is used to calculate the total number of generations which could have occurred (Table 3) (Gill 1984) to predict the numerical increase in *E. coli* (Gill *et al.* 1991).

TABLE 3.
DERIVATION OF THE BACTERIAL GROWTH NUMBER FOR A TYPICAL OFFAL
COOLING REGIME

Temp (°C)	Growth Rate (generations/h)	Time (h)	Number of generations
40–35	1.70	3.4	5.8
35–30	1.50	1.4	2.1
30–25	1.10	1.4	1.5
25–20	0.70	1.4	1.0
20–15	0.40	1.4	0.6
15–10	0.20	1.4	0.3

Bacterial growth number   11.3
Log_{10} increase   3.40
Numerical increase   2533x

Data: Gill (1984)

A three class acceptance plan with m = 9.0 generations, M = 12.0 generations, c = 20% and n = 20+ has been suggested by Gill *et al.* (1991) for the application of TFI in assessing the hygienic status of beef carcasses. The use of the three class plan would assist in setting HACCP limits for the monitoring a critical control point based on temperature and could also be used to verify the entire cooling process. In addition to carcass beef, the technique has been applied to other meats such as offals, frozen beef, poultry and fish (Olley and Ratkowsky 1973; Pooni and Mead 1984; Gill and Harrison 1985; Lowry *et al.* 1988; Gill *et al.* 1991).

Contamination Index

The contamination index estimates the levels of psychrotrophic bacteria on carcasses at different stages during processing (Gustavsson and Borch 1993). A similar system, the hygiene index, also exists but will not be discussed here (Gustavsson and Borch 1989; Gustavsson and Karleson 1989). To determine the contamination index, samples of meat from beef carcasses are taken at different processing stages. The samples are stored at 2°C for 0, 3, 7 and 14 days. The bacterial counts from plates incubated at 25°C, at each of these times, are added together, giving an 'index number' or total count/cm^2. A high index number is indicative of a high psychrotrophic count in the initial material (Gustavsson and Borch 1993). The data in Table 4 demonstrate the assessment of the hygienic status of beef carcasses using either the initial counts or the contamination index. Analysis of the data shows that the contamination index demonstrates statistically significant differences between the data for the various processing steps, while similar analysis of the initial aerobic counts does not.

TABLE 4.

ASSESSMENT OF BEEF CARCASS HYGIENE USING INITIAL AEROBIC COUNTS OR A CONTAMINATION INDEX AND THEIR VALUE AS INDICATORS OF SIGNIFICANT DIFFERENCES BETWEEN PROCESSING STAGES

Aerobic counts ($\log_{10}$ cfu/cm^2)

Processing stage	Initial[1]	t-test	Contamination[2] index	t-test
1. Before rapid chiller	2.8 (0.7)*	P<0.09a	16.6 (1.6)	P<0.03a
2. After rapid chiller and 3 h chilling	3.4 (0.7)	P<0.07b	19.5 (3.3)	P<0.98b
3. After 24 h chilling	3.0 (0.2)	P<0.31c	19.5 (1.1)	P<0.001c

*   Standard deviation
[1]   Bacterial counts on unstored meat samples
[2]   Sum of bacterial counts on meat samples stored at 2°C for 0, 3, 7 and 14 days

Comparison of processing stages:
   a   1 v 2
   b   2 v 3
   c   1 v 3

Data: Gustavsson and Borch (1993)

The contamination index can identify critical control points in HACCP programs for processes such as beef slaughter. It can also be used as a within plant reference method for more rapid on-line methods. The main disadvantage of the contamination index is that it takes weeks to obtain the data. There is an urgent need for more rapid methods to allow more rapid assessment, and better control, of bacterial numbers at critical control points.

Psychrotrophic/Psychrophilic Based Rapid Methods in HACCP Systems

Total or psychrotrophic counts, i.e., aerobic counts at 25°C can be used to show differences between the hygienic status of offal at various stages of meat processing (Sheridan and Lynch 1988). The Acridine Orange Direct Count (AODC) is a rapid method (Fig. 6) which gives an acceptable indication of psychrotrophic numbers, and takes about 10 min to complete (Sheridan et al. 1990). Log log plots of AODC and psychrotrophic counts for swabs from beef carcasses are linear with a slope of 0.99 and an $r^2 = 0.94$. Such rapid derivation of inferred psychrotrophic numbers is valuable in establishing, monitoring and verifying critical control points.

The psychrophilic count may be a better indicator than the aerobic psychrotrophic count in certain circumstances. The differences between the psychrophiles and psychrotrophs were always highly significant (P < .001) for both beef and pig carcasses stored at 4°C (Table 5). The use of the psychrophiles as indicators is further illustrated in Table 6 where different indicator organisms were used to evaluate cleaning efficiency in a beef plant. The psychrophiles were the only indicator capable of demonstrating a significant reduction in counts, as a result of cleaning. The differences between the different indicators are related to their growth characteristics. The psychrophiles are a very homogenous group, whose growth restrictions at low temperatures help define them within specific limits. This is reflected in the lower standard errors for the data in Table 6. It might be expected that the Pseudsmonads would have been good indicators but this was not the case (Table 5 and 6). The Pseudomonads are poor indicators in these experiments because they are enumerated on a selective medium and incubated at 25°C. A lack of sufficient selectivity, however, allows a wider range of bacteria to grow, and these distort the cetrimide fucidin cephaloridine (CFC) (Pseudomonad) count and lower their value as indicators.

In the future, a rapid psychrophilic method, based on the knowledge that this group is mainly Pseudomonads, may be developed. There is potential for the development of a rapid method for Pseudomonads enumeration, based on the use of specific antibodies now available (Labadie and Desnier 1993) by a development of the filtration procedure described by Sheridan et al. (1990) and the FITC procedures of Sheridan et al. (1991).

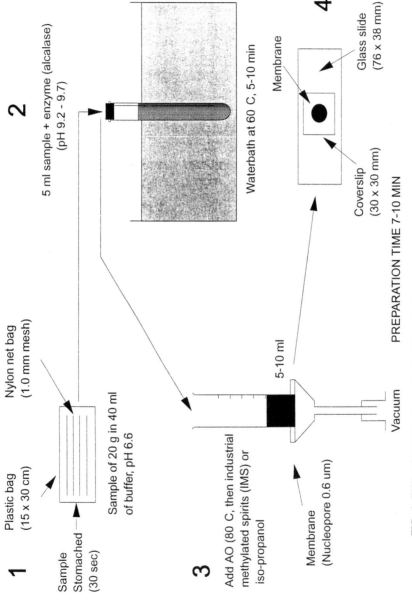

FIG. 6. THE ACRIDINE ORANGE DIRECT COUNT (AODC) TECHNIQUE FOR COUNTING BACTERIA

Data: Sheridan et al. (1990).

TABLE 5.

DIFFERENCES IN THE GROWTH OF PSYCHROTROPHS AND PSYCHROPHILES
($\log_{10}$ cfu/cm^2) ON BEEF CARCASSES STORED FOR 3 DAYS AND ON PIG
CARCASSES ASSESSED USING TWO MEDIA

| Storage | Cattle | | |
time (days)	Psychrotrophs	Psychrophiles	F-test
0	3.46	2.80	P < .001
3	3.57	3.11	
F-test	P < .05		
	Pigs		
Total counts	3.67	3.23	
			P < .001
Pseudomonads	3.04	2.83	
F-test	P < .05		
	Degrees of freedom	Standard errors of differences between means	
Cattle	179	0.12	
Pigs	76	0.15	

Psychrotrophs incubated 25°C for 3 days
Psychrophiles incubated 4°C for 14 days
Pseudomonads on CFC medium
Total count on PCA medium
Data: Sheridan (1994b)

TABLE 6.

USE OF INDICATOR ORGANISMS TO SHOW THE REDUCTION IN BACTERIAL
COUNTS ($\log_{10}$ cfu/cm^2) BEFORE AND AFTER CLEANING IN A BEEF
BONING HALL

| Surface cleaned | Stainless steel conveyor | | | | Neoprene bone conveyor | | | |
	Before	After	SED	t-test (DF4)	Before	After	SED	t-test (DF4)
Total counts[1]	4.25	4.02	0.84	NS	4.40	3.24	0.57	NS
Pseudomonads[2]	2.83	1.67	1.30	NS	2.72	1.22	0.87	NS
Psychrophiles[3]	4.24	2.85	0.51	P < .05	4.08	2.02	0.52	P < .05

[1] Aerobic counts at 25°C
[2] Counts on CFC medium at 25°C
[3] Aerobic counts at 4°C for 14 days
SED = standard error of differences between means
DF = degrees of freedom
Data: Sheridan et al. (1992)

CONCLUSIONS

Indicators may be used to assess microbiological safety, hygiene conditions during processing and product keeping quality . In this paper the indicators described are generally not indicative of the presence of specific pathogens. An exception is the temperature function integration indicator, for the growth of *E. coli*. This system determines the possible growth of *E. coli* as a pathogen itself, or of bacteria with similar growth characteristics, such as *Salmonella*. Temperature function indicators are of particular value in the operation of HACCP systems, as temperature is easily monitored on a continuous basis.

Many indictors can monitor the hygiene of the processing operations. The carcass monitoring systems, such as the contamination index, can measure the effect of different line processes, allowing effective selection of critical control points. The proposed measurements of the psychrophilic bacteria on carcasses is a further improvement of measuring process hygiene. Of particular importance for this process is the possible availability of a rapid method, which would also facilitate monitoring of critical control points on the process line (Mackey and Roberts 1993).

Apart from the total counts, the Enterobacteriaceae are also used as indicators of process hygiene. They have been used to monitor carcass hygiene during the processing of pig carcasses (Snijders 1988). They may have potential in sheep slaughter operations, but their value in monitoring carcass hygiene has been questioned because of the very low counts, resulting in a lack of proper statistical comparisons (Sierra-Castrillo *et al.* 1993).

Most chemical indicators are unsuitable for reasons of cost, complexity or a lack of sensitivity in respect of assessing microbial growth. For packaged meats, lactic and acetic acids are appropriate both in terms of speed and cost. Since these are relatively rapid tests they could be used to monitor product storage.

This paper has discussed the use of indicator systems for raw meat, particularly beef, an area of serious concern in relation to safety. Because of this concern, the introduction of HACCP systems into such operations is a priority both in the European Union and the United States (Anon. 1991; Snijders 1988; Sierra-Castrillo *et al.* 1993). Some of the available microbiological tests are inadequate, especially for monitoring critical control points. The integration of the principles underlying such systems, i.e., the use of psychrophiles, with rapid detection methods should provide more valuable and commercially relevant methods of assessing, controlling and ultimately improving meat quality and safety.

REFERENCES

ADAMCIC, M. and CLARK, D.S. 1970. Bacteria-induced biochemical changes in chicken skin stored at 5°C. J. Food Sci. *35*, 103–106.

ANON. 1991. *Methods of Enzymatic Food Analysis*, Boehringer Mannheim GmbH, Biochemica, Mannheim, Germany.

BORCH, E. and AGERHEM, H. 1992. Chemical, microbial and sensory changes during the anaerobic cold storage of beef inoculated with a homofermentative *Lactobacillus* sp. or a *Leuconostoc* sp. Int. J. Food Microbiol. *15*, 100–108.

BORTON, R.J., WEBB, N.B. and BRATZLER, L.J. 1968. The effect of microorganisms on the emulsifying capacity and extract release volume of fresh porcine tissues. Food Technol. *22*, 94–96.

BUCHANAN, R.L. 1991. Microbiological criteria for cooked, ready-to-eat shrimp and crabmeat. Food Technol. 157–160.

BUCHANAN, R.L. and BAGI, L.K. 1994. Use of aerobic plate counts incubated at elevated temperatures for detecting temperature-abused refrigerated foods: effectiveness under transitory abuse conditions. Unpublished data.

BUCHANAN, R.L., SHULTZ, F.J., GOLDEN, M.H., BAGI, K. and MARMER, B. 1992. Feasibility of using microbiological indicator assays to detect temperature abuse in refrigerated meat, poultry, and seafood products. Food Microbiol. *9*, 279–301.

CLARKE, G.C. 1989. The hazard analysis and critical point (HACCP) approach to slaughter quality control in red meat abattoirs. In *Proc. 10th Int. Symp. World Assoc. of Veterinary Food Hygienists*, Stockholm, pp. 194–198.

COLLINS, C.H. 1967. *Microbiological Methods,* Butterworth & Co., London.

DAINTY, R.H., EDWARDS, R.A., HIBBARD, C.M. and RAMANTANIS, S.V. 1986. Bacterial sources of putrescine and cadaverine in chill stored vacuum-packaged beef. J. Appl. Bacteriol. *61*, 117–123.

DAINTY, R.H. and HIBBARD, C.M. 1983. Precursors of the major end products of aerobic metabolism of *Brochothrix thermosphacta*. J. Appl. Bacteriol. *54*, 127–133.

DE PABLO, B., ASENSIO, M.A., SANZ, B. and ORDÓÑEZ, J.A. 1989. The D(-) lactic acid and acetoin/diacetyl as potential indicators of the microbial quality of vacuum-packed pork and meat products. J. Appl. Bacteriol. *66*, 185–190.

DOHERTY, A., SHERIDAN, J.J., ALLEN, P., MCDOWELL, D.A., BLAIR, I.S. and HARRINGTON, D. 1995. Survival and growth of *Aeromonas hydrophila* on modified atmosphere packaged normal and high pH lamb. Int. J. Food Microbiol. (In Press).

DORN, C.R. 1993. Review of foodborne outbreak of *Escherichia coli* O157:H7 infection in the western United States. Spec. Rep., JAVMA, *203*, 1583–1587.

EDWARDS, R.A., DAINTY, R.H. and HIBBARD, C.M. 1983. The relationship of bacterial numbers and types to diamine concentration in fresh and aerobically stored beef, pork and lamb. J. Food Technol. *18*, 777–788.

EDWARDS, R.A., DAINTY, R.H. and HIBBARD, C.M. 1985. Putrescine and cadaverine formation in vacuum packed beef. J. Appl. Bacteriol. *58*, 13–19.

FIELDS, M.L. and RICHMOND, B.S. 1968. Food quality as determined by metabolic by-products of microorganisms. Advan. Food Res. *16*, 161–229.

GARDENER, G.A. and STEWART, D.J. 1966. Changes in the free amino and other nitrogen compounds in stored beef muscle. J. Sci. Food Agric. *17*, 491–495.

GILL, C.O. 1984. Calculation of bacterial growth for evaluation of offal cooling procedures. Tech. Rep., MIRINZ 834, 1–9.

GILL, C.O. and HARRISON, J.C.L. 1985. Evaluation of the hygienic efficiency of offal cooling procedures. Food Microbiol. *2*, 63–69.

GILL, C.O., HARRISON, J.C.L. and PHILLIPS, D.M. 1991. Use of a temperature function integration technique to assess the hygienic adequacy of a beef carcass cooling process. Food Microbiol. *8*, 83–94.

GILL, C.O. and NEWTON, K.G. 1980. Development of bacterial spoilage at adipose tissue surfaces of fresh meat. Appl. Environ. Microbiol. *39*, 1076–1077.

GILL, C.O., PHILLIPS, D.M., LOEFFEN, M.P.F. and BISHOP, C. 1988. A computer program for evaluating the hygienic efficiency of meat processing procedures from product temperature history data. In *Proc. 34th Int. Meat Cong. Meat Sci. & Technol.*, Brisbane, Australia.

GUSTAVSSON, P. and BORCH, E. 1989. Contamination of beef carcasses with spoilage bacteria during slaughter and chilling. In Proc. 35th Int. Congress Meat Sci. Technol., Copenhagen, pp. 363–370.

GUSTAVSSON, P. and BORCH, E. 1993. Contamination of beef carcasses by psychrotrophic *Pseudomonas* and Enterobacteriaceae at different stages along the processing line. Int. J. Food Microbiol. *20*, 67–83.

GUSTAVSSON, P. and KARLSSON, R. 1989. Determination of critical process operations for spoilage bacteria in beef production. In World Assoc. Vet. Food Hygienists, Xth (Jubilee) Int. Symp. Stockholm, pp. 147–154.

HATHAWAY, S.C. and BULLIANS, J.A. 1992. The application of a hazard analysis critical control point system in red meat slaughter and dressing. In *Proc. World Congr. of Foodborne Infections and Intoxicants*, Berlin, pp. 895–898.

HITCHENER, B.J., EGAN, A.F. and ROGERS, P.J. 1979. Energetics of *Microbacterium thermosphactum* in glucose-limited continuous culture. Appl. Environ. Microbiol. *37*, 1047–1052.

ICMSF 1982. *Microorganisms in Foods. 2: Sampling for Microbiological Analysis: Principles and Specific applications*, University of Toronto Press.

IRMSF 1988. *Microorganisms in Foods 4. Application of the Hazard Analysis Critical Control Point (HACCP) System to Ensure Microbiological Safety and Quality,* University of Toronto Press.

JAY, J.M. 1964. Release of aqueous extracts by beef homogenates, and factors affecting release volume. Food Technol. *18*, 129–132.

JAY, J.M. 1966. Response of the extract-release volume and water-holding capacity phenomena to microbiologically spoiled beef and aged beef. Appl. Microbiol. *14*, 492–496.

JAY, J.M. 1986. Microbial spoilage indicators and metabolites. In: *Foodborne Microorganisms and Their Toxins: Developing Methodology,* (M.D. Pierson and N.J. Stern, eds.) pp. 219–240. Marcel Dekker, New York.

JAY, J.M. and KONTOU, K.S. 1964. Evaluation of the extract-release volume phenomenon as a rapid test for detecting spoilage in beef. Appl. Microbiol. *12*, 378–383.

JAY, J.M. and KONTOU, K.S. 1967. Fate of free amino acids and nucleotides in spoiling beef. Appl. Microbiol. *15*, 759–764.

KAKOURI, A. and NYCHAS, G.J. 1994. Storage of poultry meat under modified atmospheres or vacuum packs: possible role of microbial metabolites as indicator of spoilage. J. Appl. Bacteriol. *76*, 163–172.

LABADIE, J. and DESNIER, I. 1992. Selection of cell wall antigens for the rapid detection of bacteria by immunological methods. J. Appl. Bacteriol. *72*, 220–226.

LAWRIE, R.A. 1985. *Meat Science,* 4th Ed., Pergamon Press, Tarrytown, NY.

LEA, C.H., PARR, L.J. and JACKSON, H.F. 1969. Chemical and organoleptic changes in poultry meat resulting from the growth of psychrophylic spoilage bacteria at 1°C. Br. Poul. Sci. *10*, 229–238.

LOWRY, P.D., GILL, C.O. and PHAM, Q.T. 1988. A quantitative method of determining the hygienic efficiency of meat thawing processes. In *Proc. 34th Int. Cong. Meat Sci. & Technol.*, Brisbane, Australia, pp. 533–536.

MACKEY, B.M. and KERRIDGE, A.L. 1988. The effect of incubation temperature and inoculum size on growth of Salmonellae in minced beef. Int. J. Food Microbiol. *6*, 57–65.

MORTIMORE, S. and WALLACE, C. 1994. HACCP a practical approach. Chapman and Hall, London.

MOSSEL, D.A.A. and VAN NETTEN, P. 1991. Microbiological reference values for foods: a European perspective. J. Assoc. Off. Anal. Chem. *74*, 420–432.

NEWTON, K.G. 1979. Value of coliform tests for assessing meat quality. J. Appl. Bacteriol. *47*, 303-307.

NYCHAS, G.J. and ARKOUDELOS, J.S. 1990. Microbiological and physiochemical changes in minced meats under carbon dioxide, nitrogen or air at 3 °C. Int. J. Food Sci. Technol. *25*, 389-398.

OLLEY, J. and RATKOWSKY, D.A. 1973. Temperature function integration and its importance in the storage and distribution of flesh foods above the freezing point. Food Technol. Austr. *25*, 66-73.

ORDÓÑEZ, J.A., de PABLO, B., de CASTRO, B.P., ASENSIO, M.A. and SANZ, B. 1991. Selected chemical and microbiological changes in refrigerated pork stored in carbon dioxide and oxygen enriched atmospheres. J. Agric. Food Chem. *39*, 668-672.

PEARSON, D. 1968. Application of chemical methods for the assessment of beef quality. J. Sci. Food Agric. *19*, 556-559.

POONI, G.S. and MEAD, G.C. 1984. Prospective use of temperature function integration for predicting the shelf-life of non-frozen poultry-meat products. Food Microbiol. *1*, 67-78.

ROBERTS, J.A., SOCKETT, P.N. and GILL, O.N. 1989. Economic impact of a nation wide outbreak of salmonellosis: cost benefit of early intervention. Br. Med. J. *298*, 1227-1230.

ROGERS, R.E. and MCCLESKEY, C.S. 1961. Objective tests for quality of ground beef. Food Technol. *15*, 210-212.

RYAN, C.A., NICKELS, M.K. and MARGRETT-BEAN, N.T. 1987. Massive outbreak of antimicrobial-resistant salmonellosis traced to pasteurised milk. J. Am. Med. Assoc. *258*, 3269-3274.

SAFFLE, R.L., MAY, K.N., HAMID, H.A. and IRBY, J.D. 1961. Comparing three rapid methods of detecting spoilage in meat. Food Technol. 465-467.

SAYEM-EL-DAHER, N., SIMARD, R.E., FILLION, J. and ROBERGE, A.G. 1984. Extraction and determination of biogenic amines in ground beef and their relation to microbial quality. Lebensm. Wiss. Technol. *17*, 20-23.

SCHMITT, R.E. and SCHMIDT-LORENZ, W. 1992a. Formation of ammonia and amines during microbial spoilage of refrigerated broilers. Lebensm. Wiss. Technol. *25*, 6-10.

SCHMITT, R.E. and SCHMIDT-LORENZ, W. 1992b. Degradation of amino acids and protein changes during microbial spoilage of chilled unpacked and packed chicken carcasses. Lebensm. Wiss. Technol. *25*, 11-20.

SCHNEIDER, W., HILDEBRANDT, G. and SINELL, H.J. 1983. D(-) lactate concentration as a parameter for evaluating the freshness of pre-packed, heat-treated meat products. Fleischwirtschaft *63*, 1198-1205.

SHAW, B.G. and HARDING, C.D. 1984. A numerical taxonomic study of lactic acid bacteria from vacuum-packed beef, pork, lamb and bacon. J. Appl. Bacteriol. *56*, 25-40.

SHELEF, L.A. 1974. Hydration and pH of microbially spoiling beef. J. Appl. Bacteriol. *37*, 531–536.

SHERIDAN, J.J. 1994a. Unpublished data.

SHERIDAN, J.J. 1994b. Unpublished data.

SHERIDAN, J.J. and LYNCH, B. 1979. Effect of microbial contamination on the storage of beef carcasses in an Irish meat factory. Ir. J. Food Sci. Technol. *3*, 43–52.

SHERIDAN, J.J. and LYNCH, B. 1988. The influence of processing and refrigeration on the bacterial numbers on beef and sheep offals. Meat Sci. *24*, 143–150.

SHERIDAN, J.J., LYNCH, B. and HARRINGTON, D. 1992. The effect of boning and plant cleaning on the contamination of beef cuts in a commercial boning hall. Meat Sci. *32*, 185–194.

SHERIDAN, J.J., WALLS, I. and LEVETT, P.N. 1990. Development of a rapid method for enumeration of bacteria in pork mince. Irish J. Food Sci. Technol. *14*, 1–15.

SHERIDAN, J.J., WALLS, I., MCLAUCHLIN, J., MCDOWELL, D. and WELCH, R. 1991. Use of a microcolony technique combined with an indirect immunofluorescence test for the rapid detection of *Listeria* in raw meat. Lett. Appl. Microbiol. *13*, 140–144.

SIERRA-CASTRILLO, M.L., SHERIDAN, J.J. and MCGUIRE, L. 1993. The use of indicator organisms in the development of hazard analysis critical control (HACCP) system for sheep processing operations. Report to the CEC, Directorate General for Science, Research & Development, rue de la Loi 200, Brussels.

SINELL, H.J and LUKE, K. 1978. D(-) lactate as parameter for the microbial spoilage in frankfurter type sausages. In *Eur. Meeting Meat Research Workers*, Kulmbach, Germany, pp. C 11:1–C 11:6.

SLEMR, J. 1981. Biogenic amines as potential chemical indicator of meat quality. Fleischwirtschaft. *61*, 921–926.

SNIJDERS, J. 1988. Good manufacturing practices in slaughter lines. Fleischwirtschaft. *68*, 753–756.

STUDER, P., SCHMITT, R.E., GALLO, L. and SCHMIDT-LORENZ, W. 1988. Lebensm. Wiss. Technol. *21*, 224–228.

SUTHERLAND, J.P., PATTERSON, J.T. and MURRAY, J.G. 1975. Changes in the microbiology of vacuum-packaged beef. J. Appl. Bacteriol. *39*, 227–237.

TAYLOR, A.A., DOWN, N.F. and SHAW, B.G. 1990. A comparison of modified atmosphere and vacuum skin packing for the storage of red meats. Int. J. Food Sci. Technol. *25*, 89–109.

VIEHWEG, S.H., SCHMITT, R.E. and SCHMIDT-LORENZ, W. (1989). Lebensm. Wissen. Technol. *22*, 346–355.

WORTBERG, B. and WOLLER, R. 1982. Quality and freshness of meat and meat products as related to their content of biogenic amines. Fleischwirtschaft *62*, 1457–1463.

CHAPTER 8

ANIMAL HEALTH AND THE ROLE OF THE VETERINARY FOOD HYGIENIST IN THE CONTROL OF MEAT BORNE INFECTIONS

JOHN D. COLLINS

Department of Large Animal Clinical Studies
Faculty of Veterinary Medicine
University College
Dublin, Ireland

ABSTRACT

The concept of "Healthy Food from Healthy Animals" relies upon the pre- and postharvest implementation of an integrated approach for protecting the food product, from contamination or adulteration with harmful agents. This extended form of control is now an integral part of Hazard Analysis Critical Control Point Programs and provides a practical means of validating the health status of herds and flocks of food-producing animals.

In this context the veterinary food hygienist, as a food company employee, has much to offer the agri-food industry. She/He can appraise not alone the on-farm conditions but also the animal health and other data provided by the regulatory veterinarian assigned to the plant. Such information, complemented by the plant's own food quality and safety assurance program, can then be used to improve and promote the safety standards of the company. Extended health control programs emphasize the health and safety of the final product for the consumer at each stage of production and processing, and enable veterinary food hygienists, and others, to provide the agri-food industry, and governmental agencies, with a scientific basis for harmonizing trade in meat and other animal-derived foods.

INTRODUCTION

In a paper which describes the societal changes which affect the susceptibility of the human population to foodborne diseases Potter concludes with the

77

observation that "...the role of veterinarians in ensuring that our livestock and poultry are healthy, residue free and have the lowest possible level of colonization by potential foodborne pathogens will be central to any effective foodborne disease prevention program" (Potter 1992). This observation and others highlight the complexity of the food chain as it relates to the production and processing of foods of animal origin. Ultimately, the responsibility for producing a safe and wholesome food supply rests with the agri-food industry. An example of this is the success of the Irish campaign to control *Salmonella enteritidis* in the Irish poultry flock (Fallon 1994). This was attributable to the high level of cooperation and understanding between poultry producers and processors, private veterinary practitioners and the Department of Agriculture, Food & Forestry.

There is now a need, if not a requirement, to identify and address those aspects of food production, processing, distribution and retail sale which may constitute a risk to consumer health. This is not to ignore the occupational risks faced by the food producer or the hazards which arise during the "pre-harvest" phase and in the course of carcass preparation in the meat plant (Collins 1983). Rather, they are a reminder that potentially harmful biological agents exist in, or on, apparently healthy animals. These issues are addressed in this paper in the context of Hazard Analysis Critical Control Point Programs (HACCP) relating to the production and distribution of a healthy meat supply.

HACCP AND FOOD SAFETY

HACCP programs are directed towards ensuring the safety of the product for the consumer (Anon. 1993). Such programs may appear at first sight to be costly but in the final analysis may not ultimately prove to be so, when one considers the vulnerability of the industry to litigation claims. However, the introduction of "quality" parameters in such programs is cause for concern. This is because the program, extending to quality, may not be executed with the necessary degree of commitment or diligence, for reasons of cost, or for other operational reasons. The inclusion of "safety" in what becomes a quality program may be counter-productive from the consumer protection viewpoint.

RISK ANALYSIS

In attempting to quantify food-related microbiological health risks as a basis for the implementation of HACCP programs, much can be learned from the animal disease model. The risk analysis approach used in the control of the major epizootics which can affect international trade in animals and animal products is successful because it provides a scientific basis for decision making and reduces considerably the part played by "intuition" (Dahoo 1993). Risk analysis, along with its three components, risk assessment, risk management and risk communication, is now acknowledged by the Codex Alimentarius Committee of the World Health Organization, as a basis for HACCP programs in food production (Hathaway and McKenzie 1993). This ensures that when decisions are made, they are scientifically defensible even when risk is assessed on an "as low as reasonably achievable" (ALARA) principle.

This approach may offer the best means of contending with the problems posed by the emergence, over the past twenty years, of foodborne pathogens such as *Campylobacter jejuni, Escherichia coli* 0157:H7 and *Yersinia enterocolitica*. The occurrence of these species in the alimentary tract of food animals including poultry, constitutes an ever present risk even in the absence of clinical disease in the infected animals. Yet the means of addressing the risks posed by these bacteria relate directly to on-farm ("preharvest") conditions and to primary carcass dressing, and are not confined solely to the protection of the product in its final stages of processing. Much the same principle applies to the control of foodborne salmonellae of animal origin.

The provision of reliable microbiological data relating to the "preharvest" and primary processing phases, as the basis for risk analysis, is crucial to the success of HACCP programs in the meat industry. This is particularly the case when one considers the possible health-related effects of the newly introduced technologies, used to accelerate meat processing, such as *pre-rigor* deboning and accelerated maturation of pork, beef and mutton at temperatures of 20 to 25C for periods of up to five hours (Marriott 1994).

THE ROLE OF THE VETERINARIAN IN HACCP PROGRAMS

The prevention of contamination, rather than "detection inspection," is intrinsic to the implementation of HACCP programs at factory level. This is, and remains, a primary objective of the veterinary food hygienist. In achieving product protection at the "preharvest" level in integrated production systems, the HACCP principles will already have been applied by the veterinarian at several points. These include water and feed quality, animal health and performance records, animal welfare, the level of safety applied in the choice and use of therapeutic agents, and the clinical condition of the food animals, as well as the safety of the environment as affected by the standard of animal effluent management and utilization on the farm. On this basis the veterinarian can attest to the (relative) safety of the food animal and its suitability for slaughter for human consumption. Such information is a basic requirement of a truly integrated food safety assurance scheme and can readily be provided by the attending veterinarian as part of the overall program, once he/she is recognized by the food company as a full and participating member of the company's HACCP team.

The factors which determine the safety of the eventual meat product at the preharvest stage, in terms of the absence of "significant" numbers of pathogenic micro-organisms, are highly complex and require multistage logistic regression analysis, if a true weighting of risk is to be determined. Table 1 presents an example of the broad issues which require consideration in this context if presented. This approach takes into account the multi-factorial nature of the infectivity process, a process which starts with the exposure of the animal to contact with the infective agent and the progression of that agent within the

TABLE 1.
POINTS OF VETERINARY SUPERVISION AND INTERVENTION IN HACCP PROGRAMS
RELATING TO AN INTEGRATED BEEF PRODUCTION/PROCESSING SYSTEM.
EXAMPLES

1. Factors to be considered include: identification of hazard
                                           control measure(s)
                                           tolerance limits
                                           monitoring procedures
                                           corrective action
                                           person responsible

2. Levels of Production/Processing/Distribution/Usage

    (a) On-farm conditions*:   environment, animal feed, water
                                           stocking/housing diversity
                                           animal welfare/medication
                                           effluent management
                                           ANIMAL IDENTIFICATION
             *relate to multi-phase ownership of animals

    (b) Transit conditions: animal welfare
                                           level of stress
                                           direct damage
                                           intermingling of stock
                                           degree of integrity of control
                                           ANIMAL IDENTIFICATION

    (c) Meat Plant: ANIMAL IDENTIFICATION
                                           animal welfare/- lairage conditions
                                           slaughter hygiene/carcass dressing
                                           personnel competence/hygiene
                                           cold line control/product protection
                                           CARCASS/PRODUCT IDENTIFICATION
                                           dispatch
                                           HEALTH CERTIFICATION OF PRODUCT

animal body as determined by such factors as the virulence of the organism, and the level of resistance of the host animal which in turn may be compromised by malnutrition, intercurrent disease, and "stress". An example of the extent to which stress can become a factor in determining the relative risk of foodborne infection is the way in which road transportation can lead to a 10-fold increase in the numbers of campylobacters in the alimentary tract of broilers by the time they arrive at the poultry meat plant (Kazwala 1988). Another example is the extent of soiling which is seen on animals on arrival at slaughter plants; this is the result of stress induced hyper-peristalsis and undoubtedly predisposes to carcass contamination with enteric bacteria during carcass dressing. In fact the alimentary tract of the healthy animal is likely to harbor most of the bacterial and viral agents of food-borne disease (Hathaway and McKenzie 1993). Consequently any procedure which leads to contamination of the carcass and edible offals with alimentary tract contents will seriously compromise the "health" status of these foods, as shown in Table 2. The knowledge that such contamination may occur so readily and how it may be avoided, provides the means of protecting the product at this stage of processing. The practice of "bunging" the gut and tying off the esophagus are examples of how direct action at key CPP's can be highly effective.

TABLE 2.
EFFECT OF SLAUGHTER/DRESSING PROCEDURES ON THE RATE OF
CONTAMINATION OF PIG CARCASSES

a. Contamination rates with salmonellae *

		No. of carcasses examined	No. of carcasses contaminated with salmonellae
1.	Normal slaughter	35	16 (46%)
2.	Extra care at		
	- singeing		
	- evisceration	30	2 (7%)

b. Contamination rates with *Enterobacteriaceae* during evisceration **

		% of carcasses carrying > 20/sq. cm	Mean no. /sq. cm.
1.	Normal	40	60
2.	Knives, hands not washed	84	100
3.	Damage to intestines	100	1,250

*   Source: Oosterom and Notermans (1983).
**   Source: Gerats *et al.* (1981).

All factors affecting the hygiene and safety of the meat product at all stages of production, processing, and distribution are of direct concern to the regulatory veterinarian in the meat plant. When providing essential statutory certification of the health of meat products, he/she has to be satisfied, at least on an ALARA basis, (1) that the animal of origin was clinically healthy at the time of slaughter and free of specified diseases, (2) that all contaminated and diseased carcasses and parts thereof have been removed and safely disposed of, (3) that all in-plant procedure was conducted under hygienic conditions and, finally, (4) that the carcass and its offals are properly identified and otherwise described (Collins 1975). The latter requirements also apply to the transport conditions for the certified product, under European Union (EU), international and Irish law.

The effectiveness of the traditional meat inspection methods used in many developed countries in controlling and preventing foodborne disease in humans has been scrutinized in recent years (Mackey and Roberts 1991; Berends *et al.* 1993; Dubbert 1993; Hathaway and McKenzie 1993). These systems are considered to be "resource intensive" and to relate only to the detection and elimination from the food chain, of those carcasses and offals which are overtly diseased and otherwise not acceptable to the consumer. It is argued that, with the exception of, say, *Mycobacterium bovis, Erysipelothrix insidiosa, Streptococcus suis* and *Chlamydia psittaci,* few of the agents responsible for such lesions as detected at slaughter have a zoonotic potential. Nevertheless these traditional methods of veterinary examination provide a means of detecting and controlling, the beef tapeworm, *Taenia saginata,* for example, as well as providing a basis for the organoleptic assessment of, e.g., malignant tumors, abscesses, bruising and fractures, pleurisy and peritonitis, carcass pigmentation, and other abnormalities and physical defects which would be repugnant to the consumer. In these papers the authors emphasize the relatively low public health significance of the failure of affected carcasses or offals to "pass" veterinary inspection. In fact, the objective of detailed veterinary examination at the *ante-* and *postmortem* stages is to provide a reasoned basis for deciding that the majority of carcasses and offals can pass such an inspection, on a ALARA basis, and can be allowed to enter the food chain with an acceptable degree of confidence as to the "health" of the product, subject to the constraints referred to above.

Furthermore, in assessing the merits and limitations of the current carcass inspection and meat control procedures, some authors have not acknowledged the other components of regulatory veterinary intervention which provide essential support to that industry at the national level. These measures include health certification of the national herd, approval of premises for slaughter for trade, product protection through separation in plant design and operation, management, control and hygienic disposal of contaminated materials, cold line and processing control, transport hygiene and secure product identification, and health certification of product. These represent some of the components of veterinary certification of the health of the meat product and conform with the current requirements of national and international law regarding inter-state movement. Furthermore they illustrate the extent to which the basic principles of HACCP have been incorporated in Irish legislation since the 1930's (Collins 1975).

The statutory systems of health control used in meat production should be assessed periodically with a view to improvement, and to ensure that public funds are properly utilized. As the health status of the national herd improves, the contribution of detailed carcass and offal inspection for the detection of gross lesions of diseases not important to the "safety" of the product, can be modified. Governmental resources, such as veterinary personnel, could then be more effectively deployed (Dubbert 1993). In this way the HACCP approach could be made to operate more effectively as an integral part of the regulatory process in meat plants. However, it is necessary to point out that the enforcement of regulatory measures alone will not bring about all the necessary improvements in the health characteristics of the certified raw meat products, as perceived by the informed consumer. This is because, by its very nature, raw meat, i.e., red meat, poultry meat and game, is likely to harbor variable numbers of contaminating bacteria and other organisms, the presence of which, in absolute terms, is largely unavoidable. For this reason, and also because of the commercial requirements of the industry as dictated by the market place, the involvement of the industry itself in the assessment of the health acceptability of the product is a natural consequence, and is complementary to the role of the regulatory program, which it supports. In this context the veterinary food hygienist, acting in a nongovernmental capacity and as an employee of the meat industry at company level, has much to contribute.

A joint industry-regulatory agency approach based upon HACCP principles and extending back to the "preharvest" phase, is recommended. Constructive interaction between the meat industry, at company level, and the regulatory authority has much to offer both parties and can promote greater efficiency in the utilization of resources and information regarding the safety of the meat product. In this context, data collected at veterinary *postmortem* inspection, as conducted under the regulatory program, have a direct relevance to the successful operation of HACCP programs. These data provide essential information regarding, the prevalence of disease conditions in the herd or flock which may compromise the health status of the next crop of produce from that source. Direct interaction between the attending veterinarian on the farm and the meat plant/regulatory agency, can lead to effective measures being taken by the producer to control or remove any stress-related hazard thus recognized. Such intervention can also have a profound effect in controlling and avoiding residue-related problems, should they arise.

When new pathogen detection technologies are introduced, their sensitivity and specificity under commercial conditions, and the need to determine the appropriate level of expectation in regard to their usefulness in improving food safety at the different stages of production, are major concerns. In particular, concern is expressed regarding the action to be taken when a "positive" raw meat product is found, bearing in mind the need to balance the nature of the risk involved against the real commercial loss which may arise, as a consequence of actions taken on the basis of such a finding.

While the DNA- and ATP-based technologies will eventually become available for commercial use throughout the food industry, it may be some time before they replace the more established procedures. It is inevitable, however, that the new procedures will have a place in HACCP programs and will help eliminate from the food chain products which carry significant numbers of pathogenic micro-organisms. For this reason these developments are to be welcomed, but with the *caveat* that the meat industry had better be prepared to deal with the problems which the detection of unwanted microbial agents may precipitate. The HACCP approach will be enhanced by these innovations and their use will focus attention on such factors as poor plant design, ventilation and water quality, which facilitate the introduction and spread of pathogenic and spoilage microorganisms in meat plants (Hannan *et al.* 1993).

FOOD SAFETY AND THE EDUCATION OF THE CONSUMER

Despite the effectiveness of the HACCP and other approaches to the enhancement of the safety of the meat product, it is necessary to draw attention to the role of education. This applies not alone to the consumer, but also to all concerned parties, including the producer and processor and their advisers. The education of the food animal producer to reduce the exposure level of livestock to microorganisms which are of public health concern, should reduce the initial loading of the meat product with these contaminants. Coincidentally, this approach may also be essential to the successful exclusion of unwanted residues of animal remedies from the food chain, as part of the HACCP program (Collins 1985).

The Beef Quality Assurance Scheme operated in Ireland by Coras Beostoic agus Feola (CBF) places considerable emphasis upon on-farm hygiene and animal health and welfare criteria which are imposed as a condition of entry into the program. The scheme also demands full compliance with food hygiene and protection procedures which are additional to, and complement, the statutory requirements regarding plant design, operational hygiene, cold-line control, and product identification which apply to these plants for the purpose of domestic as well as international trade. Canadian studies have shown that investment in education of the food handler and end-user is highly cost-effective in the prevention and control of salmonellosis (Report 1988). Veterinarians have a central role to play in the education of the producer, processor and consumer, based on their long-standing involvement in all three phases of the food chain, and can continue to contribute, along with others, to future efforts in this area.

RESEARCH

From a veterinary viewpoint the research requirements of a successful meat safety program include:

(1) The accumulation of qualitative and quantitative microbiological data on the prevalence of pathogenic microorganisms present on, and in the environment of, food animals and their carcasses and offals at all stages of production, processing and distribution.
(2) The introduction of predictive modeling techniques in this area, based on the biological and epidemiological features of each infection or intoxication, with a view to identifying the most effective least-cost control measure(s) available.
(3) Assessment of the infective dose level of the various pathogenic agents for different consumer groups.
(4) Validation of the new microbial detection technologies, and
(5) Promotion of an integrated approach to meat safety, including drug residue avoidance, through HACCP programs.

Investment in research can be a long term process; however, it is relevant to point out that it is only in recent years, in the face of an epidemic of bovine spongiform encephalopathy (BSE) in Great Britain, that the true value of earlier research into the sheep disease, scrapie, was realized (Collee 1993). An integrated program of research is in the best interest of the agri-food industry and of the country as a whole. Data collection, applied research and directed education can contribute to the improvement of the safety of meat and meat products. Members of the veterinary profession, because of their long-standing involvement in the meat industry and the population/preventive medicine approach now commonly adopted in the practice of food animal medicine, can continue to make a considerable contribution in each of these developments.

CONCLUSION

The concept of "Healthy Food from Healthy Animals" relies upon the pre- and postharvest implementation of an integrated approach to the protection of the food product, from contamination with pathogenic and other unwanted micro-organisms and adulteration. As a primary measure the removal from the food chain of overtly diseased animals reduces the risk of animal and environmental spread of the causal agent. As a secondary measure in such cases herd health monitoring, based on an on-going analysis of herd health and production records may be useful in determining whether or not the hazard has been removed. This form of risk analysis, if adopted at farm level and with the full participation of the attending veterinarian as a member of the HACCP team, represents an "extended health control" model for safe meat production referred to above, and summarized in Table 3.

TABLE 3.
MODEL FOR EXTENDED HEALTH CONTROL OF FOODS OF ANIMAL ORIGIN

Elements include:

       Approval of production holding, water and feedstuffs
       On-farm identification and health examination of food animals*
       Veterinary certification of health of animals at farm gate
       Animal transport controls
       In-plant *ante-mortem* examination*
       Carcass identification and in-plant *postmortem* examination*
       Veterinary health certification of carcass, offals
       Cold-line/processing/transport/distribution controls*

*Including on-site screening and laboratory-based tests where applicable.

Such an approach validates the health status of herds and flocks and provide a basis for food safety assurance programs for adoption by national and international agencies and integrated food retailing chains. Today's veterinary food hygienists have much to offer in this context, as a food company employee, by providing an holistic assessment of data relating to the wholesomeness or otherwise of the food product, and by advising the company accordingly. Their professional appraisal can take into account not only on-farm conditions, but also other information provided to management by the regulatory veterinarian assigned to the plant. This includes, in the case of each individual supplier, the results of veterinary *ante-* and *post-mortem* examination of the animals and their carcasses, and also relates to the conditions of slaughtering, processing, cold line control and product protection within the packing plant and beyond, in accordance with statutory food control requirements. Such information may be complemented by the meat plant's own food quality and safety assurance program.

This HACCP-based approach enables veterinary food hygienists and others to provide both the agri-food industry, and governmental agencies, with a scientific basis for harmonizing trade in meat and other animal-derived foods. It establishes the safety of the meat product for the consumer as its central objective and, finally, when operated as part of an integrated program, it ensures that a preventive rather than a solely corrective approach can be adopted effectively at each phase of production, processing and distribution of meat and meat products.

ACKNOWLEDGEMENT

The advice and assistance of Professor John Hannan in the preparation of this paper is gratefully acknowledged.

REFERENCES

Anon. 1993. Training considerations for the application of the hazard analysis critical control point system to food processing and manufacturing. World Health Organization, Publication No. WHO/FNU/FOS/93.3. Geneva, Switzerland.

BERENDS, B.R., SNIJDERS, J.M.A. and van LOGTESTIJN, J.G. 1993. Efficacy of current EC meat inspection procedures and some proposed revisions with respect to microbiological safety: a critical review. Veterinary Rec. 133, 411–415.

COLLEE, J.G. 1993. BSE: stocktaking 1993. The Lancet 342, 790–793.

COLLINS, J.D. 1975. The veterinary food hygienist in the meat industry - present and future roles. Irish Veterinary J. 29, 83–90.

COLLINS, J.D. 1983. Abattoir-associated zoonoses. J. Soc. Occupational Med. 33, 24–27.

COLLINS, J.D. 1985. Foodborne residues and the veterinary surgeon. Irish Veterinary J. 39, 147–149.

DAHOO, I.R. 1993. Monitoring livestock health and production:service - epidemiology's last frontier? Preventive Veterinary Med. 18, 43–52.

DUBBERT, W.H. 1993. U.S. Inspection looks to the future. Proceedings of the 11th International Symposium of the World Association of Veterinary Food Hygienists (WAVFH), pp. 20–23. Bangkok, Thailand.

FALLON, M. 1994. Ireland's experience in controlling *Salmonella enteritidis*. In *Prevention and Control of Potentially Pathogenic Micro-organisms in Poultry and Poultry Meat Processing*, pp. 23–26, Proceedings of 14th meeting of FLAIR No. 6 Program held in Dublin, Ireland in February, 1994. (R.W.A.W. Mulder and J.D. Collins, eds.). Spelderholt Centre for Poultry Research and Information Services, Beekbergen, The Netherlands.

GERATS, G.E., SNIJDERS, J.M.A. and van LOGTESTIJN, J.G. 1981. Slaughter techniques and bacterial contamination of pig carcasses. Cited by Mackey and Roberts (1991). Proceedings of 27th European Meeting of Meat Research Workers, Vol. 1, pp. 198–200, Vienna, Austria.

HANNAN, J., COLLINS, J.D. and O'MAHONY, H. 1993. Factors responsible for the spread of campylobacters in poultry meat processing plants. Proceedings of the 11th International Symposium of the World Association of Veterinary Food Hygienists (WAVFH), pp. 241–244. Bangkok, Thailand.

HATHAWAY, S.C. and MCKENZIE, A.I. 1993. Risk analysis and meat hygiene. Proceedings of the 11th International Symposium of the World Association of Veterinary Food Hygienists (WAVFH), pp. 38–45, Bangkok, Thailand.

KAZWALA, R.R. 1988. In "Studies on the Origin, and Quantitative Distribution of Thermophilic Campylobacters at Various Stages of Poultry Production and Poultry Processing". M.V.M. thesis, National University of Ireland. Dublin, Ireland.

MACKEY, B.M. and ROBERTS, T.A. 1991. Hazard analysis and critical control point programmes in relation to slaughter hygiene. In *The Scientific Basis for Harmonising Trade in Red Meat*. (J. Hannan and J.D. Collins, eds.) pp. 3–18, University College Dublin. Dublin, Ireland.

MARRIOTT, N.G. 1994. Meat processing: accelerated processing of fresh meat. Meat Focus International *3*, 27–30.

OOSTEROM, J. and NOTERMANS, S. 1983. Further research into possibility of salmonella-free fattening and slaughtering of pigs. Cited by Mackey and Roberts (1991). J. Hygiene Cambridge *91*, 59–69.

POTTER, M.E. 1992. The changing face of foodborne disease. J. Am. Vet. Medical Assoc. *201*, 250–253.

Report (1988). In *Salmonellosis Control: The Role of Animal and Product Hygiene*. World Health Organisation Technical Report Series No. 744. Geneva, Switzerland.

CHAPTER 9

THE EFFECTIVENESS OF CARCASS DECONTAMINATION SYSTEMS FOR CONTROLLING THE PRESENCE OF PATHOGENS ON THE SURFACES OF MEAT ANIMAL CARCASSES[1]

GREGORY R. SIRAGUSA

United States Meat Animal Research Center
United States Department of Agriculture
Agricultural Research Service
P.O. Box 166
Clay Center, NE 68933

ABSTRACT

The effectiveness of decontamination systems for controlling the presence of pathogens and spoilage organisms on carcasses is discussed. Research using organic acids and water (hot or cold) demonstrates the general effectiveness of such treatments in lowering the aerobic plate counts (APC) on carcasses by 1–3 log_{10} cfu per unit area. Chlorine has been found to be relatively ineffective for use in animal carcass spray washers. An example of direct application of the bacteriocin nisin to inoculated tissue in a spray washer is discussed. Reductions attributable to nisin were 2–2.5 log_{10} cfu per unit area higher than those reported for various organic acids or water. Areas for further research are highlighted along with the potential use of newer technologies to elucidate attachment and detachment mechanisms of bacteria to meat animal carcasses.

INTRODUCTION

The process of converting live meat animals into carcasses for further processing has been modernized, streamlined, quickened, and made more

---

[1] Mention of a trade name, proprietary product or specific equipment is necessary to report factually on available data; however, the USDA neither guarantees nor warrants the standard of the product, and the use of the name by USDA implies no approval of the product to the exclusion of others that may also be suitable.

efficient. Despite this, all carcasses will have a microbial flora associated with them. It is not possible to reduce the level of microbial contamination on a carcass to zero using a washing system.

Carcass decontamination refers largely to the use of carcass spray washing systems which are employed to reduce and/or kill bacteria on carcasses. Most of the available research has been conducted using tissue models of the larger process of in-plant carcass washing systems. Spray washers must be judged as a single step in the whole process of producing hygienic carcasses and not as a single step towards pathogen reduction. They can only be considered a single point in a HACCP program.

The body of work presented in this review is largely from laboratory scale experiments on small pieces or cuts of meat or pilot scale carcass washers. Such research yields data that is highly variable from study to study. This variation is expected, since experiments testing carcass spray washing efficacy have a large number of variables (inoculum species, inoculum level, inoculum contact time, temperature, age of samples, enumeration assays, sampling methods, contact time, spray pressure, antimicrobial agents, concentration, etc.) which are rarely reproducible from laboratory to laboratory.

Carcass decontamination systems in the United States of America usually employ potable tap water in their sprayers. On a limited basis, some processors use low concentrations of acetic acid. This paper will review the use of organic acids, chlorine, water and recently bacteriocins, in decontamination systems designed to reduce the levels of bacterial pathogens on carcasses. The intention is to review the literature in terms of effectiveness of the practice of spray washing. For more comprehensive reviews of decontamination, references are available (Dickson and Anderson 1992).

ORGANIC ACID DECONTAMINATION

Short chain organic acids have been targeted as the most logical agents to spray on carcasses as antimicrobial agents. Lactic, acetic, citric, formic and propionic acids have all been reported in the literature for this purpose. Lactic and acetic acids are inexpensive, have GRAS status, are environmentally friendly, and are naturally occurring. Recently, longer chain fatty acid derivatives, such as monolaurin, have received attention as meat antimicrobials. Table 1 gives several examples of the reported results from using short chain organic acids as antimicrobial agents.

These data are summarized to illustrate that, although each experiment was uniquely different in design (application, inoculum, microbial detection, species, sample type, etc.) the results are remarkably similar. As a general statement, the use of organic acids reduced bacterial counts by 1–2 $\log_{10}$ cfu/area of tissue surface, regardless of the acid type. The antimicrobial effect of short chain fatty acids is largely due to lowering of the pH where the undissociated form of the acid is maintained (Gill and Newton 1982; Cutter and Siragusa 1994a). Acid concentration appears to be an important factor in the magnitude of the immediate pH drop and antimicrobial effect. However, at high acid concentrations the effects of these compounds on product quality (color, flavor) become important considerations.

The use of acid mixtures has been studied. Hypothetically, citric or ascorbic acids are thought to exert a chelating effect that enhances the pH lowering effect of either acetic or lactic acids. However this concept, while perhaps mechanistically sound, does not seem to provide a substantially larger reduction of bacterial counts on carcasses.

The last criterion for judging the effectiveness of acid spray washing is the shelf-life of stored meat products prepared from spray washed cuts. At least three studies have demonstrated no difference in the aerobic plate counts between products prepared from acid spray washed cuts and controls after refrigerated storage (Acuff *et al.* 1987; Dickson and Anderson 1991; Prasai *et al.* 1991).

CHLORINE DECONTAMINATION

The use of chlorinated water to decontaminate carcasses has received much attention (Table 2). Overall, most researchers conclude that chlorine has little or no effect, unless it is sprayed frequently over a prolonged period of time. These results are not surprising considering the infinitely larger organic load contributed by a carcass or tissue section, compared to the small amounts of free available chlorine in the spray solutions. The applied chlorine very rapidly becomes bound by the organic load and is no longer antimicrobial.

With the current climate of concern over the potential of chloramine formation from chlorine, it is probable that chlorine usage will decrease, if not be eliminated altogether. It may be replaced with compounds such as chlorine dioxide for the purposes of carcass decontamination.

WATER DECONTAMINATION AND THE ATTACHMENT PROCESS

Of all the treatments for decontaminating carcasses, rinsing the carcass with water is perhaps the single most effective means to remove at least one $\log_{10}$ APC unit from the carcass surface. In the case of hot water spray washing, the reduction is significantly higher and can be as much as >3 $\log_{10}$ APC or specific organism reduction. The data presented in Table 3 illustrates this point. The trend in this data does not differ greatly from that observed using organic acids, with the exception of hot water spray washing.

TABLE 1.
EXAMPLES OF THE USE OF SHORT CHAIN ORGANIC ACIDS TO DECONTAMINATE
MEAT ANIMAL CARCASSES OR TISSUES

Acid Type	Species	Concentration	Effect	Reference
lactic + acetic mix	lamb	6 - 24% (v/v)	$\leq 1 \log_{10}$ APC reduction	Ockerman et al. 1974
acetic + proprionic mix	swine	60:40 mix	$2 \log_{10}$ APC reduction	Reynolds and Carpenter 1974
lactic	calves	0.75 - 2.5% (v/v)	$\cong 1 \log_{10}$ APC reduction	Woolthuis and Smulders 1985
acetic, lactic	beef	1%	no significant reduction from controls after vacuum packaged storage	Acuff et. al. 1987
lactic, acetic, ascorbic citric mixture	beef	1% lactic, 2% acetic, 0.25% citric, 0.1% ascorbic	no significant reduction from controls after vacuum packaged storage	Acuff et. al. 1987
lactic	veal tongues	2% (v/v)	$\cong 3 \log_{10}$ APC reduction	Visser et al. 1988
lactic	beef	3% (v/v)	$\cong 1.8 \log_{10}$ APC reduction @ 70°C $\cong 1.1 \log_{10} E.\ coli$ reduction @ 70°C $\cong 1.2 \log_{10}$ APC reduction @ 25°C $\cong 0.4 \log_{10} E.\ coli$ reduction @ 25°C	Anderson and Marshall 1990
acetic	beef	2% (v/v) 55°C	$\geq 2 \log_{10}$ reduction of Salmonella california	Dickson and Anderson 1991
lactic	swine	1% (v/v), 55°C	$\cong 1 \log_{10}$ APC reduction	Prasai et al. 1991
lactic	beef	1% (v/v), 55°C	$\cong 1 \log_{10}$ APC reduction	Prasai et al. 1991
lactic	beef	1% (v/v), 55°C	$\cong 1.0 \log_{10}$ APC reduction immediately, but no significant difference after 3 d b/w treated and controls	Prasai et al. 1991
acetic	beef	1.5 - 3% (v/v)	$< 1.0 \log_{10}$ APC, enterobacteriaceae, and Salmonella typhimurium reduction @ 20°C	Anderson et al. 1992
lactic	beef	1.5 - 3% (v/v)	$< 1.5 \log_{10}$ APC, enterobacteriaceae, and Salmonella typhimurium reduction @ 20°C	Anderson et al. 1992

Continued

TABLE 1. (CONTINUED)

Acid Type	Species	Concentration	Effect	Reference
acetic + lactic	beef	2% acetic +1% lactic	< 1.0 $\log_{10}$ APC, enterobacteriaceae, and *Salmonella typhimurium* reduction @ 20 C	Anderson, Marshall, and Dickson, 1992
lactic + acetic	beef	2% lactic + 1% acetic	< 1.0 $\log_{10}$ enterobacteriaceae, and *Salmonella typhimurium* reduction @ 20°C $\cong$ 1.1 $\log_{10}$ APC reduction	Anderson *et al.* 1992
acetic	beef	2% (v/v) in alginate gel	$\cong$ 1.5 $\log_{10}$ *Listeria monocytogenes* reduction	Siragusa and Dickson 1992
lactic	beef	1.7% (v/v) in alginate gel	$\cong$ 1.3 $\log_{10}$ *Listeria monocytogenes* reduction	Siragusa and Dickson 1992
lactic and acetic	beef	1, 3, 5 % (v/v)	$\leq$ 1.75 $\log_{10}$ reduction of *E. coli* O157:H7	Cutter and Siragusa 1994a
lactic and acetic	beef	1, 3, 5 % (v/v)	$\cong$ 3.5 $\log_{10}$ reduction of *Pseudomonas fluorescens*	Cutter and Siragusa 1994a

The fact that water spraying can often effect a comparable reduction in APC, as spraying with organic acids, is evidence that the physical removal of bacteria from the layer of surface water surrounding the carcass is the important step in effective spray washing. The process of bacterial attachment is complex even if the underlying attachment surface is inert and homogenous (e.g., stainless steel). In the case of an animal carcass, the number of variables increases greatly and the attachment process becomes orders of magnitude more complicated. The attachment process is thought to be in two stages (Marshall 1977), the first stage being reversible and the second stage more permanent or irreversible. The first stage is largely characterized as a physical phenomenon between the particle (bacterium) and the surface (carcass). The second stage is thought to be accompanied by the synthesis or action of either a general attachment substance (e.g., extracellular bacterial polysaccharide) or a specific attachment to substrate entities (e.g., lectins).

General antimicrobial agents, such as organic acids, are apparently effective on those organisms that are contacted by the acid for a long enough period of time to effect a lethal or inhibitory response. However, for organisms that are not in contact with the agent for a sufficient period of time, or have become entrapped or protected in the carcass surface, the application of an antimicrobial has no real effect. In the literature, the concentrations of organic acids applied are often well below the levels needed to effect any adverse physiological response, much less lethality, for species such as *E. coli* (Cutter and Siragusa 1994a). Therefore, the combination of a means to physically remove or loosen attached bacteria on carcasses followed by an antimicrobial spray or hot water,

TABLE 2.

EXAMPLES OF THE USE OF CHLORINE TO DECONTAMINATE MEAT ANIMAL CARCASSES OR TISSUES

Agent	Species	Concentration	Effect	Reference
chlorine	beef	200 mg/l	$\leq 2 \log_{10}$ APC immediate reduction $\geq 2 \log_{10}$ APC reduction post 24 h	Kotula et al. 1974
chlorine	beef	100 - 400 mg/l	1.5 - 1.8 $\log_{10}$ APC reduction post 24h	Emswiler et al. 1974
chlorine	beef	electrically generated	no significant APC reductions between treated and untreated	Marshall et al. 1977
chlorine	beef	200 mg/l	no differences in APC, coliforms, or staphylococci between treated and untreated samples	Stevenson et al. 1978
chlorine	beef	200 mg/l	initial reduction in APC but no differences post 8 d	Titus et al. 1978
chlorine	beef	200 mg/l	no differences in APC or total lactics between treated and untreated	Johnson et al. 1979
chlorine	lamb	450 mg/l @ 80°C	immediate APC reduction but no differences post 7d	Kelly et al. 1981
chlorine	pork	200 mg/l	1.5 $\log_{10}$ psychrotrophic count reduction	Skelly et al. 1985
chlorine	beef	50 - 800 mg/l	no significant E. coli reductions between treated and untreated	Cutter and Siragusa 1995

addresses the two steps (i.e., detachment and inactivation) necessary for a carcass decontamination system to be effective.

Most of the model systems used to evaluate the effectiveness of carcass washing systems use a very large level of inoculum or rely on endogenous contamination to demonstrate spray washing efficacy. Both of these extremes can lead to erroneous conclusions. Researchers could benefit by using a more moderate or realistic level of organisms inoculated on tissue ($\log_{10}$ 2–3 cfu/cm^2). As an example, Barkate et al. (1993) sprayed hot water on uninoculated carcasses that had bacterial levels of $\log_{10}$ 2.3–2.4 APC cfu/cm^2 before treatment. Following treatment with hot water (95°C at source, 82°C on carcass surface) the levels decreased to a mean of $\log_{10}$ 1.3 APC cfu/cm^2. The levels detected following treatment are at the lower limits of detection for the standard plate count. These results are probably a more accurate reflection of commercial carcass washer efficacy than laboratory model generated data with artificially inoculated carcass tissues.

TABLE 3.
EXAMPLES OF THE USE OF WATER RINSING TO DECONTAMINATE
MEAT ANIMAL CARCASSES OR TISSUES

Agent	Species	Conditions	Effect	Reference
water	beef	85 - 499 kPa pressure	significant surface reduction b/w pressures	Kotula et al. 1974
water	beef	80°C	2.64 $\log_{10}$ E. coli reduction 2.21 $\log_{10}$ Salmonella reduction	Smith and Graham 1978
water	mutton	80°C	3.32 $\log_{10}$ E. coli reduction 3.33 $\log_{10}$ Salmonella reduction	Smith and Graham 1978
water	beef	commercial carcass washing unit	1.07 $\log_{10}$ APC reduction	Anderson et al. 1987
water	beef	hand washed with hose	0.99 $\log_{10}$ APC reduction	Anderson et al. 1987
water	beef	83.5°C at carcass surface	2.23 $\log_{10}$ APC reduction, 10 sec exposure; 2.98 $\log_{10}$ APC reduction, 20 sec exposure	Davey and Smith 1989
water	beef	550 kPa (80 psi)	1 - 1.9 $\log_{10}$ reduction of E. coli O157:H7	Cutter and Siragusa 1994b
water	beef	95°C spray gum application, 82°C at carcass surface	1.05 $\log_{10}$ APC reduction on uninoculated beef sides	Barkate et al. 1993

BACTERIOCINS AS CARCASS DECONTAMINATING AGENTS

To date, the use of bacteriocins as carcass decontaminating agents has been reported only once (Cutter and Siragusa 1994b). Nisin was applied to separate beef carcass surface tissues, inoculated with approximately $\log_{10}$ 4 cfu/cm^2 of *Listeria innocua, Carnobacterium divergens* or *Brochothrix thermosphacta*. Reductions of 3.3, 3.0 and 3.6 $\log_{10}$ cfu/cm^2 respectively, were effected after storage of carcasses for 24 h at 5°C. The control tissues sprayed with water showed a reduction of < 1 $\log_{10}$ for each species. The magnitude of the nisin reduction is largely dependent on the susceptibility of the individual strain and species of target organism. A major step would be to demonstrate the effectiveness of bacteriocins against Gram negative pathogens. A measure of the effect of bacteriocin purity in eliminating pathogens would also be valuable. This will require new or different agents with wider bactericidal range or the use of agents (i.e., EDTA) that enhance the current efficacy of bacteriocins. The application of bacteriocinogenic cultures directly to the carcass is another possibility.

CONCLUSIONS AND FUTURE QUESTIONS TO BE ADDRESSED

For the removal of gross physical contaminants and the improvement of carcass appearance, spray washing has a definite role. The microbiological effects of spray washing are limited, irrespective of the substance used. Reductions of 1–2 $\log_{10}$ cfu/cm^2 can be achieved with these treatments. The reduction in bacterial counts is usually immediate in the case of organic acid treatments, and efficacy should be determined on a longer term basis. When considering this factor, the limited data suggests that from a shelf-life perspective, spray washing with organic acids has no lasting effects. However, in relation to pathogen removal, at least one report demonstrates that the levels of pathogens on acid treated tissue are lower than on untreated tissue. The production of safe carcasses at an economical treatment cost will also be a priority.

An understanding of the interaction of bacteria with the surface water layer of carcasses and the mechanisms of bacterial attachment will be pivotal if carcass washing treatments are to be made more effective in pathogen removal or inactivation. Since the levels of pathogens on carcasses are generally low, spray washing might have a positive public health effect (Dickson and Siragusa 1994).

Bacteriocins show great potential for use in carcass decontamination, especially for specific pathogens, although their use will require approval as a food additive. Ultimately, cost may be the deciding factor in their usage as carcass decontaminating agents.

Altering the final microfloral population could have potential negative effects on what are considered to be beneficial bacteria, such as lactics, or spoilage pseudomonads that are potential organoleptic signals of spoilage. Where all viable bacteria (benign, beneficial or pathogenic) on a carcass surface are eliminated, very real safety concerns arise because the potential for pathogen dominance becomes possible. If the goal of carcass decontamination is to lower the frequency of pathogen transmission, this is achievable when these systems are used as a part of the whole process of pathogen reduction.

REFERENCES

ACUFF, G.R., VANDERZANT, C., SAVELL, J.W., JONES, D.K., GRIFFIN, D.B. and EHLERS, J.G. 1987. Effect of acid decontamination of beef subprimal cuts on the microbiological and sensory characteristics of steaks. Meat Sci. *19*, 217–226.

ANDERSON, M.E., HUFF, H.E., NAUMANN, H.D., MARSHALL, R.T., DAMARE, J.M., PRATT, M. and JOHNSTON, R. 1987. Evaluation of an automated beef carcass washing and sanitizing system under production conditions. J. Food Prot. *50*, 562–566.

ANDERSON, M.E. and MARSHALL, R.T. 1990. Reducing microbial populations on beef tissues: concentration and temperature of lactic acid. J. Food Safety *10*, 181–190.

ANDERSON, M.E., MARSHALL, R.T. and DICKSON, J.S. 1992. Efficacies of acetic, lactic, and two mixed acids in reducing numbers of bacteria on surfaces of lean meat. J. Food Safety *12*, 139–147.

BARKATE, M.L., ACUFF, G.R., LUCIA, L.M. and HALE, D.S. 1993. Hot water decontamination of beef carcasses for reduction of initial bacterial numbers. Meat Sci. *35*, 397–401.

CROUSE, J.D., ANDERSON, M.E. and NAUMANN, H.D. 1988. Microbial decontamination and weight of carcass beef as effected by automated washing pressure and length and time of spray. J. Food Prot. *51*, 471–474.

CUTTER, C.N. and SIRAGUSA, G.R. 1994a. Efficacy of organic acids against *Escherichia coli* O157:H7 attached to beef carcass tissue using a pilot scale model carcass washer. J. Food Prot. *57*, 97–103.

CUTTER, C.N. and SIRAGUSA, G.R. 1994b. Decontamination of beef carcass tissue with nisin using a pilot scale model carcass washer. Food Microbiol. *11*, 481–489.

CUTTER, C.N. and SIRAGUSA, G.R. 1995. Application of chlorine to reduce populations *Escherichia coli* on beef. J. Food Safety *15* (1), 67–75.

DAVEY, K.R. and SMITH, M.G. 1989. A laboratory evaluation of a novel hot water cabinet for the decontamination of sides of beef. Int. J. Food Sci. Technol. *24*, 305–316.

DICKSON, J.S. and ANDERSON, M.E. 1991. Control of *Salmonella* on beef tissue surfaces in a model system by pre- and post-evisceration washing and sanitizing, with and without spray chilling. J. Food Prot. *54*, 514–518.

DICKSON, J.S. and ANDERSON, M.E. 1992. Microbiological decontamination of food animal carcasses by washing and sanitizing systems: A review. J. Food Prot. *55*, 133–140.

DICKSON, J.S. and SIRAGUSA, G.R. 1994. Survival of *Salmonella typhimurium, Escherichia coli* O157:H7, and *Listeria monocytogenes* during storage on beef sanitized with organic acids. J. Food Safety *14*, 313–327.

EMSWILER, B.S., KOTULA, A.W. and ROUGH, D.K. 1976. Bactericidal effectiveness of three chlorine sources used in beef carcass washing. J. Anim. Sci. *42*, 1445–1450.

GILL, C.O. and NEWTON, K.G. 1982. Effect of lactic acid concentration on growth on meat of gram-negative psychrotrophs from a meat-works. Appl. Environ. Microbiol. *43*, 284–288.

JOHNSON, M.G., TITUS, T.C., McCASKILL, L.H. and ACTON, J.C. 1979. Bacterial counts on surfaces of carcasses and in ground beef from carcasses sprayed or not sprayed with hypochlorous acid. J. Food Sci. *44*, 169–173.

KELLY, C.A., DEMPSTER, J.F. and McLOUGHLIN, A.J. 1981. The effect of temperature, pressure, and chlorine concentration of spray washing water on numbers of bacteria on lamb carcasses. J. Appl. Bact. *51*, 415–424.

KOTULA, A.W., LUSBY, W.R., CROUSE, J.D. and de VRIES, B. 1974. Beef carcass washing to reduce bacterial contamination. J. Anim. Sci. *39*, 674–679.

MARSHALL, R.T., ANDERSON, M.E., NAUMANN, H.D. and STRINGER, W.C. 1977. Experiments in sanitizing beef with sodium hypochlorite. J. Food Prot. *40*, 246–249.

OCKERMAN, H.W., BORTON, R.J., CAHILL, V.R., PARRETT, N.A. and HOFFMAN, H.D. 1974. Use of acetic and lactic acid to control the quantity of microorganisms on lamb carcasses. J. Milk Food Technol. *37*, 203–204.

PRASAI, R.K., ACUFF, G.R., LUCIA, L.M., HALE, D.S., SAVELL, J.W. and MORGAN, J.B. 1991. Microbiological effects of acid decontamination of beef carcasses at various locations in processing. J. Food Prot. *54*, 868–872.

REYNOLDS, A.E. and CARPENTER, J.A. 1974. Bactericidal properties of acetic acids on pork carcasses. J. Anim. Sci. *38*, 515–519.

SIRAGUSA, G.R. and DICKSON, J.S. 1992. Inhibition of *Listeria monocytogenes* on beef tissue by application of organic acids immobilized in a calcium alginate gel. J. Food Sci. *57*, 293–296.

SKELLY, G.C., FANDINO, G.E., HAIGLER, J.H. and SHERARD, R.C. 1985. Bacteriology and weight loss of pork carcasses treated with a sodium hypochlorite solution. J. Food Prot. *48*, 578–581.

SMITH, G.C., VARNADORE, W.L., CARPENTER, Z.L. and CALHOUN, M.C. 1976. Post-mortem treatment effects on lamb shrinkage, bacterial counts, and palatability. J. Anim. Sci. *42*, 1167–1174.

SMITH, M.G. and GRAHAM, A. 1978. Destruction of *Escherichia coli* and salmonellae on mutton carcasses by treatment with hot water. Meat Sci. *2*, 119–128.

STEVENSON, K.E., MERKEL, R.A. and LEE, H.C. 1978. Effects of chilling rate, carcass fatness and chlorine spray on microbiological quality and case life of beef. J. Food Sci. *43*, 849–852.

TITUS, T.C., ACTON, J.C., McCASKILL, L. and JOHNSON, M.G. 1978. Microbial persistence on inoculated beef plates sprayed with hypochlorite solutions. J. Food Prot. *41*, 606–612.

VISSER, I.J.R., KOLMEES, P.A. and BIJKER, P.G.H. 1988. Microbiological conditions and keeping quality of veal tongues as affected by lactic acid decontamination and vacuum packaging. J. Food Prot. *51*, 208–213.

WOOLTHUIS, C.H.J. and SMULDERS, F.J.M. 1985. Microbial decontamination of calf carcasses by lactic acid sprays. J. Food Prot. *48*, 832–837.

CHAPTER 10

USE OF IRRADIATION TO KILL ENTERIC PATHOGENS ON MEAT AND POULTRY[1]

D.W. THAYER

US Department of Agriculture
Agricultural Research Service
Eastern Regional Research Center
Food Safety Research Unit
600 East Mermaid Lane
Philadelphia, PA 19118

ABSTRACT

Ionizing radiation can be an effective step in a HACCP program to kill enteric pathogens associated with meat and poultry products. The populations of most common enteric pathogens such as Campylobacter jejuni, Escherichia coli *O157:H7,* Staphylococcus aureus, Salmonella spp., Listeria monocytogenes, *and* Aeromonas hydrophila *can be significantly decreased or eliminated by low-dose (< 3.0 kGy) treatments with ionizing radiation. Only the enteric viruses and the endospores of the genera* Clostridium *and* Bacillus *are highly resistant to ionizing radiation, and even these are affected to some degree. Temperature effects must be carefully considered, as reduced irradiation temperatures not only result in fewer adverse changes in the sensorial properties of meat and poultry products, but they also may demand that greater radiation doses be used to inactivate the foodborne pathogen. Equations have been developed that predict the effects of low-dose ionizing radiation treatments on several enteric pathogens associated with meat and poultry. In some cases substrate-specific differences in radiation resistance have been found. Irradiation in combination with vacuum packaging or modified-atmosphere packaging has increased both safety and shelf-life for some types of refrigerated products. High-dose irradiation treatments of enzyme-inactivated meat or poultry products in vacuo at subfreezing temperatures can be used to prepare sterile, shelf-stable products with excellent sensorial properties. Such products have been used extensively by military organizations, NASA during space flights, yachtsman, and hospitals. Contemporary research will be reviewed.*

---

[1] Mention of brand or firm names does not constitute an endorsement by the US Department of Agriculture over others of a similar nature not mentioned.

INTRODUCTION

Ionizing radiation can kill enteric pathogens associated with meat and poultry products. The populations of most common enteric pathogens associated with meat products can be significantly decreased or eliminated by low-dose ($<$ 3.0 kGy) treatments with ionizing radiation. The radiation resistance of some of these foodborne pathogens is compared in Table 1. *Campylobacter jejuni* is the most common pathogen on poultry products and is very sensitive to treatment with gamma radiation (Lambert and Maxcy 1984). Current United States regulations require a minimum dose of 1.5 kGy and a maximum dose of 3.0 kGy for the irradiation of poultry. The minimum dose should inactivate 10^9 colony forming units (CFU) of *Campylobacter*, between 10^2 and 10^4 CFU of *Salmonella*, and 10^4 CFU of *Staphylococcus*. *Listeria* is more resistant to gamma radiation than most serovars of *Salmonella*; however, a dose of 1.5 kGy should reduce the viable population by at least 10^2 CFU and a dose of 2.5 kGy should inactivate at least 10^3 CFU of this pathogen. Lewis and Corry (1991) reported that the numbers of *Listeria* positive chicken carcasses are significantly lower following irradiation treatment at 2.5 kGy. The lag phase of *L. monocytogenes* at 6°C following a radiation dose of 2.5 kGy is extended from 1 day to 18 days (Patterson *et al.* 1993). This means that low levels of pathogen survival would not be a problem during the product's shelf-life.

TABLE 1.
CONTROL OF ENTERIC PATHOGENS WITH IONIZING RADIATION

Pathogen	Temp. (°C)	Substrate	D Value kGy	Reference
Aeromonas hydrophila	2	Beef	0.14 - 0.19	Palumbo *et al.* (1986)
Campylobacter jejuni	0-5	Beef	0.16	Lambert and Maxcy (1984)
Escherichia coli O157:H7	5	Beef	0.28	Thayer and Boyd (1993)
Listeria monocytogenes	2-4	Chicken	0.77	Huhtanen *et al.* (1989)
Salmonella spp.	2	Chicken	0.36 - 0.77	Thayer *et al.* (1990)
Staphylococcus aureus	0	Chicken	0.36	Thayer and Boyd (1992)

These predictions, however, assume that the radiation sensitivities of the pathogens will be very similar on meat and poultry products. Equal radiation sensitivities may not occur with all pathogens on different meats or poultry products. It is well established that pathogens such as *Salmonella* when suspended in buffer, broth, or mechanically deboned chicken (Thayer *et al.* 1990) and *L. monocytogenes* when it is suspended in buffered saline or poultry meat (Patterson 1989) have different radiation sensitivities.

The value of treating poultry and red meats with ionizing radiation to control *Salmonella* is widely reported. A radiation dose of 6 kGy reduces the population of *Salmonella* on frozen deboned horse or kangaroo meat by a factor of at least 10^5 (Ley *et al.* 1970). Gamma irradiation is an effective method to control *Salmonella* on poultry carcasses (Licciardello *et al.* 1970). Mulder (1976) estimated from results from irradiation of frozen poultry carcasses artificially contaminated with *S. panama* that a dose of 7 kGy would be necessary to achieve a contamination level of 1 in 10,000 carcasses or lower. The population of naturally occurring *Salmonella* on deep frozen or fresh chicken carcasses is reduced by $10^{2.5}$ by a dose of 2.5 kGy (Mulder *et al.* 1977). *Salmonella* populations continue to decline during cold storage ($-18°C$) of irradiated (2.5 kGy) chicken (Mulder 1982). *Salmonella* can effectively be controlled by irradiation of beef (Mossel *et al.* 1972; Mossel 1977; Tarkowski *et al.* 1984a,b). Grant and Patterson (1992) investigated the sensitivity of *Clostridium perfringens, S. typhimurium, L. monocytogenes, S. aureus,* and *Bacillus cereus* to gamma radiation on the components (beef, gravy, cauliflower, roast potato, and mashed potato) of a ready-to-eat meal and found that *C. perfringens, L. monocytogenes,* and *S. typhimurium* vegetative cells have similar ranges of D values (0.3 - 0.7 kGy) on most of the substrates. The D values in gravy were significantly less than those on the other substrates. The vegetative cells of *B. cereus* had the lowest D value of the five pathogens.

Elevated irradiation temperatures significantly reduce the survival of *Salmonella* and subfreezing temperatures during irradiation increase survival (Licciardello 1964; Previte *et al.* 1970). *Salmonella typhimurium* cells surviving low doses of gamma radiation are more sensitive to heat than nonirradiated cells, and thus are unlikely to survive even a mild cooking process (Thayer *et al.* 1991). This effect persists for at least six weeks during refrigerated storage of the irradiated chicken (Fig. 1).

Contamination of ground beef in the United States with *E. coli* O157:H7 caused 477 cases of severe hemorrhagic diarrhea, some of which progressed into hemolytic uremic syndrome (HUS) with three deaths (Anon. 1993a,b). This pathogen is associated primarily with undercooked and raw beef, lamb, pork, and poultry (Doyle and Schoeni 1987; Neil 1989; Snyder 1992). *E. coli* O157:H7 was the cause of a large (243 cases, 4 deaths) waterborne outbreak of hemorrhagic diarrhea in Burdine Township, Missouri, during December of 1989 and January of 1991 indicating that it may be more widespread than previously suspected (Swerdlow *et al.* 1992). Thus, it was appropriate to investigate its sensitivity to gamma radiation when it contaminated either beef or poultry. The presence or absence of air does not significantly alter the survival of *E. coli* O157:H7 on mechanically deboned chicken meat and lean ground beef when it was gamma irradiated (Thayer and Boyd 1993). The temperature of irradiation, however, significantly affects the survival of *E. coli* O157:H7, and its response

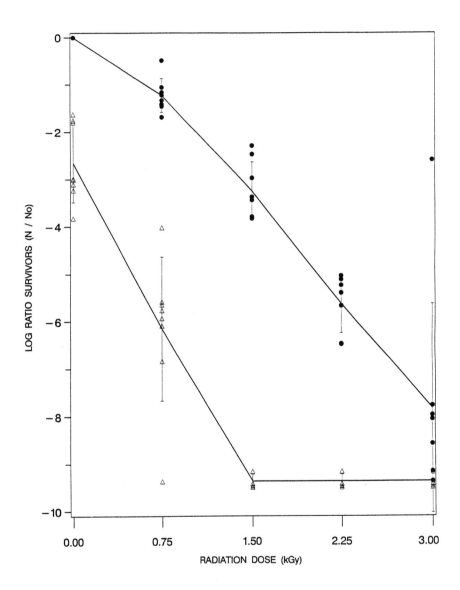

FIG. 1. RESPONSE OF *S. TYPHIMURIUM* IN VACUUM-PACKAGED, MECHANICALLY
DEBONED CHICKEN MEAT TO TREATMENT WITH GAMMA RADIATION AT 0°C (●)
OR TO TREATMENT WITH GAMMA RADIATION AT 0°C FOLLOWED BY HEATING
AT 60°C FOR 2.0 MIN (Δ)
Samples were heated and analyzed after 0, 2, 4, and 6 weeks of storage at 5°C.
All results are presented. The vertical bar represents the geometric mean ±SD.
Adopted from Thayer *et al.* 1991.

is illustrative of responses that will be encountered with other microorganisms as well. We discovered that the radiation resistance of *E. coli* O157:H7 is much greater below freezing than is typical of other foodborne pathogens. In Figure 2, its survival following exposure to a radiation dose of 1.5 kGy *in vacuo* is plotted against the temperature of irradiation from -60 to +15°C. The very

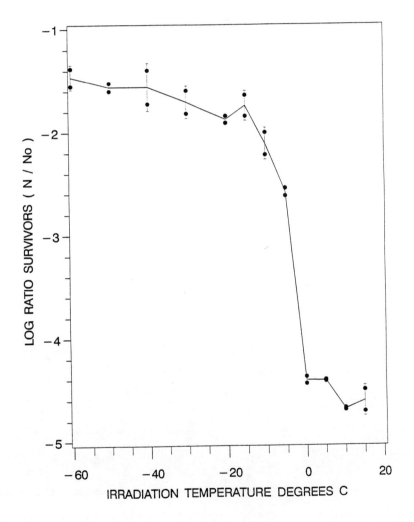

FIG. 2. RESPONSE OF *E. COLI* O157:H7 IN FINELY GROUND LEAN BEEF TO A DOSE OF 1.5 KGY WHEN IRRADIATED *IN VACUO* AT TEMPERATURES OF -60 TO +15°C
Adopted from Thayer and Boyd (1993).

sharp changes in its survival around the freezing point led us to determine the
D-values in ground lean beef at temperatures of ±5°C (Fig. 3). The radiation
resistance of *E. coli* O157:H7 was nearly identical on beef and chicken. These
regressions are equivalent to D values of 0.28 ± 0.02 kGy at +5°C versus 0.44
± 0.03 kGy at -5°C. By statistically analyzing the results of a response surface

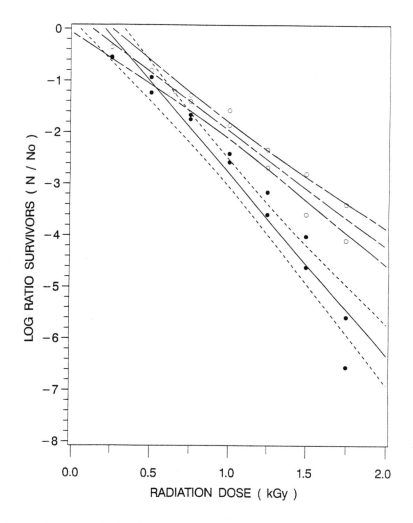

FIG. 3. RESPONSE OF *E. COLI* O157:H7 WHEN IRRADIATED IN FINELY GROUND
LEAN BEEF AT +5 OR -5°C *IN VACUO*
Closed circles, +5°C; open circles, -5°C; dashed lines, 95% confidence limits on
mean predicted values. Adopted from Thayer and Boyd (1993).

study of the survival of *E. coli* O157:H7 on beef irradiated at temperatures from -20 to +20°C to doses from 0 to 3.0 kGy we developed the following equation, which can be used to predict the effects of various combinations of irradiation temperature and dose:

$$\text{Log}_{10}\text{survivors} = -0.6410 - 1.289 \text{ (kGy)} + 0.002 \text{ (temperature)} - 0.070 \text{ (kGy)}$$
$$\text{(temperature)} - 1.301 \text{ (kGy)}^2 + 0.002 \text{ (temperature)}^2 \text{ } (R^2 = 0.931)$$

These results are presented graphically in Fig. 4. We have developed similar predictive equations for the responses of *Salmonella* and *Staphylococcus* on chicken to gamma radiation at various temperatures (Thayer and Boyd 1991a,b, 1992). A challenge study was conducted using an inoculum of $10^{4.8}$ CFU/g of lean beef in which neither viable CFU nor toxin was found in samples following irradiation to 1.5 kGy and then temperature abused at 35°C for 20 h (Table 2). Irradiation is an effective method to control this organism in a HACCP program.

C. botulinum would not be substantially inactivated by radiation doses below 3 kGy. Thus, current USDA and U.S. FDA regulations require that oxygen-permeable packaging materials be used for irradiated poultry products because substantial reductions of the normal indigenous microflora might otherwise allow the undetected multiplication and toxin production by *C. botulinum*. Studies of chicken skins inoculated with *C. botulinum* Type E and incubated both aerobically and anaerobically at an abuse temperature of 10°C revealed that the natural flora multiplies and produces spoilage odors within 8 days, and *C. botulinum* type E spores surviving a dose of 3 kGy do not produce toxin within 14 days (Firstenberg-Eden *et al.* 1982, 1983). Chicken skins inoculated with spores of *C. botulinum* types A and B produced similar results (Dezfulian and Bartlett 1987).

There are similar concerns about the possibility of rapid multiplication of other pathogens such as *Salmonella* in meats contaminated either before or after processing. Szczawiska *et al.* (1991) discovered that constant inocula of uninjured *Salmonella dublin, S. enteritidis,* and *S. typhimurium* added to mechanically deboned chicken meat that had received radiation doses of 0, 1.25, or 2.50 kGy prior to inoculation multiplied only slightly faster in samples incubated aerobically at either 10 or 20°C.

An area of emerging importance is the combination of various forms of modified atmosphere packaging (MAP) with irradiation treatments. Combination treatments have the potential of achieving both substantially increased shelf-life and microbiological safety (Thayer 1993). For example, Grant and Patterson (1991) investigated the microbiological safety of minced pork inoculated with *S. typhimurium, L. monocytogenes, E. coli, Yersinia enterocolitica,* and *C. perfringens* packed in 25% CO_2 and 75% N_2 and then irradiated and subjected

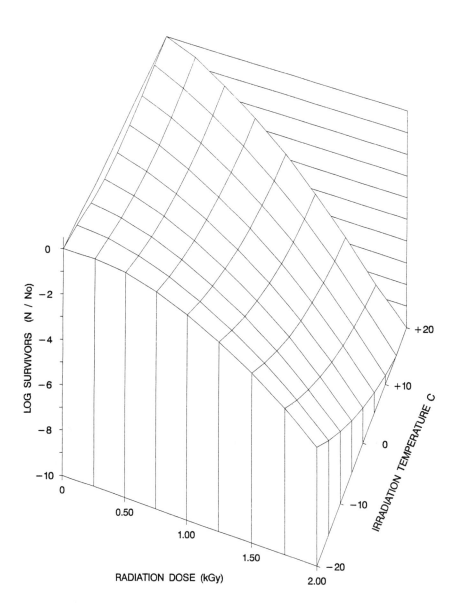

FIG. 4.   PREDICTED SURVIVAL OF *E. COLI* O157:H7 IN FINELY GROUND LEAN
BEEF FOLLOWING GAMMA IRRADIATION AT TEMPERATURES OF -20°C TO +20°C.
Data adopted from Thayer and Boyd (1993).

TABLE 2.
E. *COLI* O157:H7 SURVIVING TREATMENT WITH GAMMA RADIATION ON LEAN
GROUND BEEF AND PRODUCTION OF VERO TOXIN DURING SUBSEQUENT
TEMPERATURE ABUSE

Storage (°C)	Assay	0 kGy	0.75 kGy	1.50 kGy	2.25 kGy	3.0 kGy
5	CFU/g	5.6×10^4	Not Detected[a]	Not Detected	Not Detected	Not Detected
5	Verotoxin[b]	Negative	Negative	Negative	Negative	Negative
35	CFU/g	1.0×10^8	1.6×10^4	Not Detected	Not Detected	Not Detected
35	Verotoxin	2389	3	Negative	Negative	Negative

[a]Not Detected: < 10 CFU/g.
[b]50% cytotoxic doses.
Results are the average of three independent experiments.
Adopted from Thayer and Boyd (1993).

to temperature abuse during storage. Irradiation to 1.75 kGy reduces pathogen levels to below the detectable limit. Spoilage is apparent when either spoilage or pathogenic microorganisms reach unacceptably high numbers. They concluded that the microbiological safety of the combination process is better than MAP alone.

High-dose irradiation treatments of enzyme-inactivated meat or poultry products *in vacuo* at sub-freezing temperatures can be used to prepare sterile, shelf-stable products with excellent sensorial properties. A wide variety of irradiated shelf-stable products are used extensively by military organizations, NASA during space flights, yachtsman, and hospitals. NASA has regularly used irradiated shelf-stable beefsteak, corned beef, turkey slices, ham, and pork sausage because of their safety and lack of requirements for refrigeration (Josephson 1983). Such products should be especially attractive for use by hospitals and by consumers requiring minimal exposure to foodborne pathogens (Aker 1984).

Low-dose treatments of poultry or red meats can be used as an effective intervention step in HACCP programs to control foodborne pathogens. Irradiated poultry and fruit are available at competitive prices in the United States and are well received by the public (Pszczola 1983).

REFERENCES

AKER, S. N. 1984. On the cutting edge of dietetic science. Nutrition Today. *19*, (4) 24–27.

Anon. 1993a. Preliminary report - Foodborne outbreak of *Escherichia coli* O157:H7 infections from Hamburgers - Western United States. Morbidity Mortality Reports. *42*, 85–86.

Anon. 1993b. Update: Multistate outbreak of *Escherichia coli* O157:H7 infections from hamburgers -- Western United States, 1992-1993. Morbidity Mortality Reports. *42*, 258-263.

DEZFULIAN, M. and BARTLETT, J.G. 1987. Effects of irradiation on growth and toxigenicity of *Clostridium botulinum* types A and B inoculated onto chicken skins. Appl. Environ. Microbiol. *53*, 201-203.

DOYLE, M.P. and SCHOENI, J.L. 1987. Isolation of *Escherichia coli* O157:H7 from retail fresh meats and poultry. Appl. Environ. Microbiol. *53*, 2394-2396.

FIRSTENBERG-EDEN, R., ROWLEY, D.B. and SHATTUCK, G.E. 1982. Factors affecting growth and toxin production by *Clostridium botulinum* type E on irradiated (0.3 Mrad) chicken skins. J. Food Sci. *36*, 186-197.

FIRSTENBERG-EDEN, R., ROWLEY, D.B. and SHATTUCK, G.E. 1983. Competitive growth of chicken skin microflora and *Clostridium botulinum* Type E after an irradiation dose of 0.3 Mrad. J. Food Prot. *46*, 12-15.

GRANT, I.R. and PATTERSON, M.F. 1991. Effect of irradiation and modified atmosphere packaging on the microbiological safety of minced pork stored under temperature abuse conditions. Int. J. Food Sci. Technol. *26*, 521-533.

GRANT, I.R. and PATTERSON, M.F. 1992. Sensitivity of foodborne pathogens to irradiation in the components of chilled ready meal. Food Microbiol. *9*, 95-103.

HUHTANEN, C.N., JENKINS, R.K. and THAYER, D.W. 1989. Gamma radiation sensitivity of *Listeria monocytogenes*. J. Food Prot. *52*, 610-613.

JOSEPHSON, E.S. 1983. Radapertization of meat, poultry, finfish, shellfish and special diets. In *Preservation of Food by Ionizing Radiation* (E.S. Josephson and M.S. Peterson, eds.) pp. 231-251, CRC Press, Boca Raton.

LAMBERT, J.B. and MAXCY, R.B. 1984. Effect of gamma radiation on *Campylobacter jejuni*. J. Food Sci. *49*, 665-667, 674.

LEWIS, S.J. and CORRY, J.E.L. 1991. Survey of the incidence of *Listeria monocytogenes* and other *Listeria* spp. in experimentally irradiated and in matched unirradiated raw chickens. Int. J. Food Microbiol. *12*, 257-262.

LEY, G.J., KENNEDY, T.S., KAYASHIMA, K., ROBERTS, D. and HOBBS, B.C. 1970. The use of gamma radiation for the elimination of *Salmonella* from frozen meat. J. Hyg. Camb. *68*, 293-311.

LICCIARDELLO, J.J. 1964. Effect of temperature on radiosensitivity of *Salmonella typhimurium*. J. Food Sci. *29*, 469-474.

LICCIARDELLO, J.J., NICKERSON, J.T.R. and GOLDBLITH, S.A. 1970. Inactivation of *Salmonella* in poultry with gamma radiation. Poultry Sci. *49*, 663-675.

MOSSEL, D.A.A. 1977. The elimination of enteric bacterial pathogens from food and feed of animal origin by gamma irradiation with particular reference to *Salmonella* radicidation. J. Food Qual. *1*, 85-104.

MOSSEL, D.A.A., KROL, B. and MOERMAN, P.C. 1972. Bacteriological and quality perspectives of *Salmonella* radicidation of frozen boneless meats. Alimenta. *2*, 51–60.

MULDER, R.W.A.W. 1976. Radiation inactivation of *Salmonella panama* and *Escherichia coli* K 12 present on deep-frozen broiler carcasses. Eur. J. Appl. Microbiol. *3*, 63–69.

MULDER, R.W.A.W. 1982. The use of low temperatures and radiation to destroy Enterobacteriaceae and Salmonellae in broiler carcasses. J. Fd. Technol. *17*, 461–466.

MULDER, R.W.A.W., NOTERMANS, S. and KAMPELMACHER, E.H. 1977. Inactivation of salmonellae on chilled and deep frozen broiler carcasses by irradiation. J. Appl. Bacteriol. *42*, 179–185.

NEIL, M.A. 1989. *E. coli* O157:H7--current concepts and future prospects. J. Food Safety *10*, 99–106.

PALUMBO, S.A., JENKINS, R.K., BUCHANAN, R.L. and THAYER, D.W. 1986. Determination of irradiation D-values for *Aeromonas hydrophila*. J. Food Prot. *49*, 189–191.

PATTERSON, M. 1989. Sensitivity of *Listeria monocytogenes* to irradiation on poultry meat and in phosphate-buffered saline. Lett. Appl. Microbiol. *8*, 181–184.

PATTERSON, M.F., DAMOGLOU, A.P. and BUICK, R.K. 1993. Effects of irradiation dose and storage temperature on the growth of *Listeria monocytogenes* on poultry meat. Food Microbiol. *10*, 197–203.

PREVITE, J.J., CHANG, Y. and EL-BISI, H.M. 1970. Effects of radiation pasteurization on *Salmonella*. I. Parameters affecting survival and recovery from chicken. Can. J. Microbiol. *16*, 465–471.

PSZCZOLA, D. 1993. Irradiated poultry makes U.S. debut in midwest and Florida markets. Food Technol. *47*(11), 89–96.

SNYDER, O.P. Jr. 1992. HACCP--an industry food safety self-control program, part III. Dairy Food Environ. Sanit. *12* (3), 164–167.

SWERDLOW, D.L., WOODRUFF, B.A., BRADY, R.C., GRIFFIN, P.M., TIPPEN, S., DONNELL, H.D., GELDREICH, E., PAYNE, B.J., MEYER, A., WELLS, J.G., GREENE, K.D., BRIGHT, M., BEAN, N.H. and BLAKE, P.A. 1992. A waterborne outbreak in Missouri of *Escherichia coli* O157:H7 associated with bloody diarrhea and death. Ann. Internal Med. *117*, 812–819.

SZCZAWISKA, M.E., THAYER, D.W. and PHILLIPS, J.G. 1991. Fate of unirradiated *Salmonella* in irradiated mechanically deboned chicken meat. Int. J. Food Microbiol. *14*, 313–324.

TARKOWSKI, J.A., STOFFER, S.C.C., BEUMER, R.R. and KAMPELMACHER, E.H. 1984a. Low dose gamma irradiation of raw meat. I. Bacteriological and sensory quality effects in artificially contaminated samples. Int. J. Food Microbiol. *1*, 13–23.

TARKOWSKI, J.A., BEUMER, R.R. and KAMPELMACHER, E.H. 1984b. Low gamma irradiation of raw meat. II. Bacteriological effects on samples from butcheries. Int. J. Food Microbiol. *1*, 25–31.

THAYER, D.W. 1993. Extending shelf-life of poultry and red meat by irradiation processing. J. Food Prot. *56*, 831–833, 846.

THAYER, D.W. and BOYD, G. 1991a. Effect of ionizing radiation dose, temperature, and atmosphere on the survival of *Salmonella typhimurium* in sterile, mechanically deboned chicken meat. Poultry Sci. *70*, 381–388.

THAYER, D.W. and BOYD, G. 1991b. Survival of *Salmonella typhimurium* ATCC 14028 on the surface of chicken legs or in mechanically deboned chicken meat gamma irradiated in air or in vacuum at temperatures of -20 to +20°C. Poultry Sci. *70*, 1026–1033.

THAYER, D.W. and BOYD, G. 1992. Gamma ray processing to destroy *Staphylococcus aureus* in mechanically deboned chicken meat. J. Food Sci. *57*, 848–851.

THAYER, D.W. and BOYD, G. 1993. Elimination of *Escherichia coli* O157:H7 in meats by gamma irradiation. Appl. Environ. Microbiol. *59*, 1030–1034.

THAYER, D.W., BOYD, G., MULLER, W.S., LIPSON, C.A., HAYNE, W.C. and BAER, S.H. 1990. Radiation resistance of *Salmonella*. J. Indust. Microbiol. *5*, 383–390.

THAYER, D.W., SONGPRASERTCHAI, S. and BOYD, G. 1991. Effects of heat and ionizing radiation on *Salmonella typhimurium* in mechanically deboned chicken meat. J. Food Prot. *54*, 718–724.

CHAPTER 11

USE OF TIME/TEMPERATURE INTEGRATORS, PREDICTIVE MICROBIOLOGY, AND RELATED TECHNOLOGIES FOR ASSESSING THE EXTENT AND IMPACT OF TEMPERATURE ABUSE ON MEAT AND POULTRY PRODUCTS

THEODORE P. LABUZA and BIN FU

University of Minnesota
Department of Food Science and Nutrition
St. Paul, MN 55108

ABSTRACT

Temperature abuse is a main concern for chilled and/or MAP meat and poultry products since it will not only cause economic loss, but may also lead to a foodborne illness hazard. A major question for such products is whether organoleptic spoilage due to microbial or chemical action will occur before pathogen numbers or toxin levels become a risk when a product undergoes abuse temperatures. Time-temperature integrators (TTI) have shown potential to indicate or estimate temperature abuse. However knowledge of predictive food microbiology is needed for each food to employ a TTI tag. The simple Monod and the nonlinear Gompertz function can be used to obtain growth parameters (i.e., lag time, exponential growth rate) from kinetic studies of microbial spoilage. The temperature dependence of these parameters is best evaluated by the Arrhenius or the square root model. Effects of temperature on other deteriorative processes can also be described similarly. Three approaches are presented to evaluate the extent and impact of temperature abuse on meat and poultry by use of TTI tags, which include the effective temperature, the equivalent time and the equivalent point approach. Other technologies to evaluate or to reduce the impact of temperature abuse are also discussed.

111

INTRODUCTION

Meat and poultry products have frequently been implicated in outbreaks of foodborne illness. The estimated annual health care and personal costs, because of injury due to a few selected foodborne pathogens in 1992, was about $5–$6 billion, while about $4 billion in costs have been attributed to meat and poultry products (Anon. 1993). Most common errors contributing to foodborne disease involve some aspect of time-temperature control (Tompkin 1990). Temperature control in the distribution channel is critical in maintaining the quality and safety of chilled products. The optimum range for successfully handling and displaying refrigerated foods is −1 to 2°C (30–35°F), certainly never higher than 5°C (40°F). Unfortunately, the existing distribution channel is not well equipped for the optimum control of temperature during the distribution and display of refrigerated foods. Temperature abuse is common throughout the distribution and retail markets, with the temperature in 21% of household refrigerators often higher than 10°C. Recent data suggested that 33% of retail refrigerated foods were held in display cases above 7°C and 5% were held above 13°C (Anon. 1989). Temperatures were even higher in southern market regions. Serious microbial stability problems exist because of the frequency of temperature abuse (Doyle 1991; Almonacid-Merino and Torres 1993). The remaining shelf-life of a refrigerated food is largely determined at any given time by the cumulative effect of temperature during the previous handling of the product. Almonacid-Merino and Torres (1993) did a simulation study to evaluate the consequences of temperature abuse and showed that, even when the fraction of the total storage time at an undesirable condition (e.g., room temperature) is rather small (2–3%), the reduction in shelf-life can be highly significant (20–30%).

Traditionally, a large portion of meat and poultry products is sold frozen. In addition, fresh meat and poultry are cut and packaged locally, and are placed into a freezer once the consumer brings them home from the store. Freezing accompanied by proper thawing and cooking before consumption should lead to a palatable and safe food. However, the recent trend to manufacture fresh foods that last for weeks, or months in the refrigerator, i.e., made with "invisible" processing methods, which are not perceived as processing by the consumer, creates a new paradigm shift for food safety control. The use of modified atmosphere packaging (MAP) technology to extend the shelf-life of perishable products, through manipulation of the oxygen and carbon dioxide levels in the shipping container or package, has been with us for a long time but is now being used more extensively. In addition new active packaging technologies are coming to the forefront to create extended shelf-life refrigerated (ESLR) products (Labuza et al. 1992, Labuza 1994). Thus, the problem of temperature abuse for meat and poultry will become more critical.

The safety and quality of MAP foods depends on product characteristics, processing and packaging procedures, and environmental conditions during distribution and storage. A major question for chilled and/or MAP meat and poultry products is whether organoleptic spoilage due to chemical or microbial action will occur before the pathogen numbers or toxin levels become a risk when a product undergoes cycling or abuse temperatures. High numbers of pathogens increase the probability of foodborne diseases with the same cooking process and if heat resistant toxins are produced, cooking will not solve the problem. There are several possible ways to alleviate this problem: (1) application of the Hazard Analysis and Critical Control Point (HACCP) system during handling, processing and distribution (Corlett 1989), an approach under study by the USDA; (2) optimization of modified atmospheric (MA) conditions to contribute to the inhibition of pathogens; (3) incorporation of new, and possibly controversial, processing steps, such as chlorine washing or low dose ionizing irradiation (Lambert *et al.* 1991a) (4) use of active packaging technologies with the specific aim of inhibiting microbial (pathogen) growth, e.g., ethanol emitters or an antibacterial film around the food (Labuza 1993, 1994); and (5) inoculation with lactic acid bacteria or other competitors to control *Clostridium botulinum* outgrowth through either production of natural anti-microbial agents such as lactic acid and nisin or through biological competition in refrigerated foods or minimally processed products such as sous-vide foods (Gombas 1989). All of these means warrant further investigation to ensure the safety of fresh foods packaged under modified atmospheres, especially when exposed to temperature-abuse storage conditions. Table 1 lists the potential methods for shelf-life extension of meat or poultry, some of which fit into the "invisible processing" category.

As noted in Table 1, temperature is a primary factor for maintenance of the safety and quality of refrigerated meat and poultry products. Use of the other factors listed in Table 1 allows for some synergistic effects in shelf-life extension, while pasteurization reduces the initial load but may selectively eliminate competitive spoilage organisms. Temperature abuse may occur at any point in the meat and poultry distribution chain and jeopardize the product safety or shelf-life. The possibility of temperature abuse or lack of control in distribution, resulting in potential pathogen growth or toxin production, has limited the popularity of MAP meat and poultry in the USA (Labuza 1993). Perhaps the most practical approach is to (1) develop an effective cold distribution chain for chilled and/or MAP meat and poultry, which requires the maintenance of low storage temperature throughout distribution and minimizes temperature abuse in order to prevent pathogen growth, and (2) to ship, sell and consume the product within a short period of time, as is done in Europe and Japan. In short, attention should be paid to the third part of the HACCP approach, i.e., on the distribution chain. As shown earlier, with the limitations of the current refrigeration systems and the poor control in the distribution chain, temperature fluctuations currently are unavoidable. The questions are how can these fluctuations be minimized or how much fluctuation is practically acceptable? Significant progress has been made recently to answer the second question through the use of simple and inexpensive indicators for the detection and evaluation of temperature abuse (Taoukis et al. 1991).

A time-temperature indicator or integrator (TTI) is a small self-adhesive tag or label that can indicate temperature abuse or keep track of an accumulated time-temperature distribution function to which a perishable product is subjected from the point of manufacture to the display shelf of the retail outlet if on the case, or eventually to the consumer if on the individual food package. Generally, TTIs can be used for monitoring the cold distribution system, improving inventory management, assisting open dating of perishable foods, and ensuring food safety (Taoukis et al. 1991). Reviews and studies on TTI applications for a variety of food products have been done by several researchers (Fields and Prusik 1986; Wells and Singh 1988a).

This paper mainly presents the potential applications of TTIs for estimating the extent and impact of temperature abuse in the distribution chain and for use in monitoring the safe shelf-life of chilled and/or MAP meat and poultry products based on predictive microbial kinetics. Other related technologies to estimate or reduce the impact of temperature abuse or to improve shelf-life will also be discussed briefly.

TABLE 1.
SAFETY BARRIERS FOR SHELF-LIFE EXTENSION OF REFRIGERATED FOODS[a]

Type	Barrier
Primary	Refrigeration ($<4°C$)
Secondary	$a_W < 0.91$
	pH < 4.6
	High levels of nonpathogenic competing organisms
	$\geq$ 120 ppm nitrite (meat or poultry products)
	NaCl 2-3.5% (meat or poultry products)
	CO_2 ($>$ 10 to 20%)
	$O_2 \cong 2\%$ (CAP/MAP products) [b]
	antimicrobial agents (natural or synthetic)
	scavenger/emitter/active packaging
	mild pasteurization (heat, μ-waves, irradiation, light)

[a]Modified from AFDO (1991)
[b]From Hintlian and Hotchkiss (1986)

PREDICTIVE MICROBIOLOGY

In order to specify an appropriate time-temperature indicator for a product, the most important information needed is the product's shelf-life data. The end point of shelf-life for chilled and/or MAP meat and poultry products is generally determined by organoleptic quality, microbial spoilage and microbial safety (Labuza 1982). Sensory perceptions (e.g., meat color), evidence of metabolic by-products and types and levels of microorganisms are all valuable, and together give a full picture of food quality and safety. These food quality deterioration processes follow either pseudo-first order (e.g., microbial growth) or pseudo-zero order kinetics (Labuza 1984). Spoilage of meat and poultry by microbial growth usually requires a high number of cells to be present. The dominant organisms leading to spoilage of a particular product depend on the sources and the environmental conditions under which the product is stored. For example, in refrigerated packaged beef, *Pseudomonas* spp. were dominant aerobically while *Lactobacillus* spp. and others were dominant anaerobically (Zamora and Zaritzky 1985).

Predictive microbiology offers an alternative to traditional microbiological evaluation of food quality and safety. The concept is that detailed knowledge of the microbial ecology of a food product can be expressed as a mathematical model to estimate the effect of processing, storage and distribution processes on microbial development (Labuza *et al.* 1992; Buchanan 1993). A model could be more accurate than the data used to build it because it amplifies hidden patterns and discards unwanted noise (Gauch 1993). Predictive microbiology can be applied to determine and predict the shelf-life of perishable foods under commercial conditions based on microbial growth kinetics. It can also be used to predict the safety of a product which lacks inhibitors to microbial growth. Predictions based on mathematical models for microbial growth and safety are being applied to the distribution of chilled foods (Pooni and Mead 1984; Labuza *et al.* 1992; McMeekin *et al.* 1992).

Determination of Growth Kinetic Parameters

A general microbial growth curve on a semilog plot includes a lag phase, an exponential phase, and a stationary phase, each of which can be modeled by either linear or nonlinear regression procedures. A death phase may also be included if inactivation of microbes during cooking and cooling is of interest. There is little agreement among microbiologists regarding the best mathematical model to apply. The general Monod (1949) doubling time equation has a kinetic basis, and it is used extensively in biochemical engineering. It states that the rate at which the population increases is proportional to the number of members in the population, i.e., the specific growth rate or the generation time is assumed to be constant for constant environmental conditions. The integrated form of the Monod model is written as:

$$N = N_o \exp[k(t-t_L)] \tag{1}$$

where N is the number of organisms at time t, N_o is the initial number, k is the specific growth rate, and t_L is the lag time. The specific growth rate is the slope of the plot of ln N vs. t (or log N vs. T and k = 2.303 × slope), for $t > t_L$. The lag time, t_L, can be determined graphically. This model is simple and fairly accurate and has been used extensively but does not account for the lag phase, an important factor in shelf-life extension.

Alternatively, several nonlinear regression models have been proposed to describe a microbial growth curve. A current model used by many predictive microbiologists is the Gompertz function, which was first introduced by Gibson *et al.* (1987) to this area. Due to space and nutrient limitation, as well as toxic metabolite production, the specific growth rate of microbes actually is not constant over time, but increases to a maximum, then decreases. The Gompertz function has the form:

$$\text{LogN} = A + C \exp \{-\exp[-B (t-M)]\} \tag{2}$$

where: N is the number of organisms at time t, A = asymptotic log count of bacteria as time decreases indefinitely, C = asymptotic amount of growth that occurs as t increases indefinitely (i.e., number of log cycles of growth), M = time at which the absolute growth rate is maximal, and B = relative growth rate at M.

Lag phase duration, exponential growth rate or generation time as well as maximum population density can be derived from the above four parameters by non-linear regression of log N vs. time (Gibson *et al.* 1987; Buchanan and Phillips 1990) or can be estimated directly from its modified form (Zwietering *et al.* 1990). Both the Gompertz function and its modified form have been tested for many species of microorganisms and fit the data very well (Buchanan and Phillips 1990; Willcox *et al.* 1993). However, the fit of the Gompertz function is greatly affected by the number of observations made for the growth curve and the statistical quality of those counts (Bratchell *et al.* 1989). Both the Gompertz function and Monod equation are designed for constant temperature, but both contain a temperature dependent term, i.e., the growth rate constant, thus prediction for other temperatures can be made as long as an adequate temperature model is available.

Arrhenius Temperature Dependence Model

The Arrhenius model, long used as the standard temperature dependence model for reaction rates can also be used to describe the temperature sensitivity of microbial growth where the rate of growth is exponentially proportional to the inverse of the absolute temperature by:

$$k = k_o \exp [-E_A/(RT)] \tag{3}$$

In this equation, k_A is the rate constant at a given temperature T, k_o is the preexponential factor, E_A is the activation energy of the microbial growth, R is the universal gas constant and T is the temperature (K). Temperature sensitivity of other quality deterioration processes in meat and poultry can also be described by the Arrhenius equation (Labuza 1982). Labuza (1983) has shown that the commonly used Q_{10} term (the change in reaction rate for a 10°C rise in temperature), is directly related to the E_a and temperature range of interest as will be shown subsequently. For meat and poultry, in general the Q_{10} can be from about 2 to as much as 15 fold. This temperature effect points out the misconception of using the area under the time temperature history curve as an index of quality loss or microbial growth, i.e. if the Q_{10} was 4 and the relative growth rate was 1 at 0°C, meat held for 1 day at 20°C (rate of $4 \times 4 \times 1$ and area of $20 \times 1 = 20$) would have a relative growth increase of 16 while meat held for 10 days at 2°C (rate of $\sim 5 \times 10$ and area of $10 \times 2 = 20$), i.e. the same area under the T vs. t curve, would have a relative growth of 50 times, over three- fold more growth than the short term higher temperature abuse exposure. This shows the need for a model to integrate the data for the T/t variable.

Combining Eq. (1) and (3), one may estimate the microbial level at any other constant temperature after a certain period as long as k_o and E_a are known:

$$\log_e (N/N_o) = k_o \exp [-E_A/(RT)] (t-t_L) \tag{4}$$

To do this, tests using a minimum of three temperatures are needed, but five temperatures improve the predictions significantly. From this one can also estimate the microbial level in a product if exposed to a variable time-temperature history, where the log change in cfu is:

$$\log_e (N/N_o) = k_o \int \{\exp[-E_A/(RT_t)]\} dt \tag{5}$$

In the Eq. (5), T_t represents temperature as a function of time.

Lag time is very important for chilled and/or MAP meat and poultry, since it is an significant portion of shelf-life. The Arrhenius relationship can also be applied to model the temperature dependence of the lag phase. The inverse of the lag time is used in constructing the Arrhenius plot (i.e., $\log t_L$ vs. $1/T$ gives a straight line).

The activation energy values of microbial growth usually range between 15–40 Kcal/mol (Q_{10} of 2 to 15 at refrigerated temperature) (Labuza et al. 1992), which can be affected by other factors, such as a_w, O_2 or CO_2 concentration. For example, from the data for shelf-life of cut-up chicken by Ogilvy and Ayres (1951), the calculated E_A values for bacterial growth are 24.8 Kcal/mol for 0% CO_2, 28.2 Kcal/mol for 15% CO_2, and 31.9 Kcal/mol for 25% CO_2. This implies that the microbial growth rate is more temperature sensitive at higher CO_2 levels.

For pathogenic organisms, the end point for acceptability is essentially equal to the minimum detection level of the organism, which is based on the analytical method used. However, the presence of low numbers of certain pathogens (e.g., *Staphylococcus aureus*, *Clostridium perfringens*, *Bacillus cereus*) may not be hazardous and thus some action level must be set. For example, *S. aureus* at 10 cfu/g may be acceptable for many foods. In extended storage, toxin may be produced if there is a toxin-producing pathogen present; then the shortest (not the mean) lag time before the toxin can be detected at any growth condition should be used as the end of the shelf-life (Baker and Genigeorgis 1990).

Square Root Temperature Dependence Model

Ratkowsky *et al.* (1982) proposed a simple two parameter empirical equation for the temperature dependence of microbial growth up to the optimum temperature (T_{opt}) as:

$$\sqrt{k} = b(T - T_{min}) \qquad (6)$$

where k is the specific growth rate from the growth curve as before, b is the slope of the regression line of $\sqrt{k}$ vs. temperature, T is the test temperature (in either C or K) and T_{min} is the notational microbial growth temperature where the regression line cuts the temperature axis at $\sqrt{k} = 0$. Based on statistical data transformation, this model fits most of the microbial growth rate data very well (Ratkowsky *et al.* 1982, 1983). Since almost all TTIs will show the Arrhenius type relationship, as will be seen later, this model will not be used to correlate microbial growth in food with the TTI.

Ratkowsky *et al.* (1983) expanded this model to cover up to the maximum growth temperature, where:

$$\sqrt{k} = b(T - T_{min}) \{1 - \exp[c(T - T_{max})]\} \qquad (7)$$

where b is the regression coefficient of $\sqrt{k}$ vs. T for $T < T_{opt}$, c is an additional parameter to enable the model to fit the data for temperatures above the T_{opt}, and T_{max} is the upper temperature where the regression line intersects the temperature axis at $\sqrt{k} = 0$. This model will be useful to estimate the effect of cooking or cooling process on microbial inactivation in the food preparation stage.

Exponential Temperature Dependence Model

If the temperature range is small (~ 20 to $30°C$), then the specific growth rate or the time to a specific population level plotted on a semi-log paper versus temperature is also a straight line. The exponential model takes the form of:

$$k = 1/t_{sl} = k_o exp(bT) \qquad (8)$$

where k is the growth rate at temperature T in C, t_{sl} is the time to a specific growth level, k_o is the growth rate at $0°C$, and b is the slope. This model is also applicable to lag phase data or shelf-life data (Labuza 1984) and in many cases also used to correlate with a TTI.

As previously noted, another term often used is Q_{10}, which is defined as the ratio of rate constants or shelf-lives at temperatures differing by $10°C$. The activation energy, b in Eq. (8), and Q_{10} are interrelated through the following equation:

$$lnQ_{10} = 10b = \frac{10E_A}{RT(T+10)} \qquad (9)$$

There are many other temperature-dependent models for microbial growth in the literature (Zwietering *et al.* 1991; Labuza *et al.* 1992). However, in general, they have more parameters to be estimated, which makes it very difficult to use for prediction under variable temperature conditions and almost impossible to employ a TTI tag.

TIME/TEMPERATURE INTEGRATORS

Types of TTI

The principles of TTI operations are either a mechanical, chemical or enzymatic irreversible change usually expressed as a visible response in the form of a mechanical deformation, color development or color movement. The visible response gives information on the storage conditions to which the tag has been exposed. The LifeLines Fresh-Scan® (Fields and Prusik 1986) is based on the solid state polymerization of a thinly-coated colorless acetylenic monomer that changes to a highly-colored opaque polymer. The measurable change is the reflectance, which is measured with a laser optic wand. The data can be read and stored in a handheld device. The indicator has two bar codes, one for identification of the product and the other for identification of the indicator model. The indicators have to be stored in the freezer prior to use since they are active from the time of production.

Consumer-readable TTIs function on the same principles as continuous response indicators, but are simpler and lower in cost. Their end points can be visually recognized by ordinary consumers. Attached on individually packaged products, the tag can serve as dynamic or active shelf-life labeling instead of, or in conjunction with, open-date labeling. The TTIs assure the consumers that the products have been properly handled and indicate remaining shelf-life. An example is Lifeline's FreshCheck as shown in Fig. 1 and used on a number of meat products, including frozen turkey rolls and meat salads and luncheon meats.

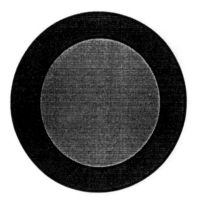

**Use by May 1992 Unless
inner circle is darker than outer ring**

FIG. 1. LIFELINES FRESH-CHECK* CONSUMER-READABLE INDICATOR

The 3M company (St. Paul, MN. USA) offers a TTI system, called MonitorMark™ (Manske 1983). This is based on a time-temperature dependent diffusion of a dyed fatty acid ester along a porous wick made of a high quality blotting paper as shown in Fig. 2. The measurable response is the distance of the advancing diffusion front from the origin. Before use, the blue dye/ester mixture is separated from the wick by a barrier film so that no diffusion occurs. To activate the indicator, the barrier is pulled off and diffusion starts if the temperature is above the melting point of the ester which determines the response temperature of a particular tag. Once activated, migration will only occur when the temperature is above the tag's response temperature. Thus, this type of TTI can be called a critical temperature/time integrator (CTTI). Another type of TTI made by the 3M is the dual temperature indicator (DTI), which has a second, higher temperature set point on a CTTI. These dual temperature tags provide users with a greater degree of information, recording exposure and duration at one temperature level, as well as an indication of temperature abuse at a second higher level. The 3M company also makes critical temperature indicators (CTIs) that show exposure above a critical temperature (T_c) after a certain period (a few minutes up to a few hours), but do not show the history of exposure above or below the temperature. They merely indicate that the product was exposed to an undesirable temperature for a short period of time sufficient to cause a change critical to its safety or quality. An example of such an application would be a tag for the indication of *C. botulinum* growth which occurs only above the critical temperature of 3.3°C (Notermans *et al.* 1990).

The I-point® TTI (Blixt 1983) is based on a color change caused by a pH decrease due to a controlled enzymatic hydrolysis of a lipid substrate. The company recently has restarted operations after having financial difficulties. Before activation the pancreatic lipase and the lipid substrate (tricaproin) are in two separate compartments. The TTI is triggered with a special activating device and can be applied manually or mechanically depending on the packaging line. At activation, the barrier that separated the compartments is broken, the enzyme and the substrate are mixed, the pH drops and the color change starts, shown by the presence of an added pH indicator. The color change can be visually recognized or instrumentally measured.

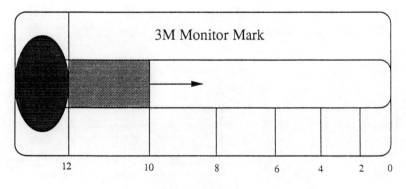

Months remaining good for use at 2 - 8 C

FIG. 2. 3M MONITORMARK™ DIFFUSED-BASED INDICATOR

Interpretation of Output of TTIs

Several kinetically based correlation approaches have been developed based on the Arrhenius temperature dependence model (Fu and Labuza 1993). The $E_{A(TTI)}$ values of the indicators cover the range of the most important deteriorative reactions in foods. Table 2 lists the activation energy and Q_{10} values of various types of TTIs (Fu and Labuza 1992).

To calculate the cumulative effects of a variable time-temperature history, the concept of an effective temperature (T_{eff}) is introduced. T_{eff} is defined as the constant temperature that results in the same quality change as the variable temperature distribution over the same period of time. Thus the effective rate constant (k_{eff}) at the T_{eff} can be expressed as:

$$k_{eff(food)} = k_{oA} \exp\left(-\frac{E_A(food)}{RT_{eff}(food)}\right) \qquad (10)$$

TABLE 2.
VALUES OF ACTIVATION ENERGY AND Q_{10} FOR SEVERAL TTIs

Producer	Model of TTI	E_A (kcal/mol)	Q_{10}
LifeLines	18[a]	27.0	5.8
	21[b]	21.3	4.0
	34[b]	17.8	3-2
	41[a]	20.5	3.8
	57[b]	21.3	4.0
	68[a]	19.7	3.6
	A20[c]	19.4	3.5
	A40[c]	19.5	3.6
3M	4P[a][*]	9.8	1.9
I-point	2090[d]	13.1	2.3
	2180[b]	14.3	2.5
	2220[b]	14.0	2.5
	3014[c]	11.4	2.1
	3270[d]	21.1	4.0
	4004[d]	40.0	13.5
	4007[a]	32.7	8.4
	4014[c]	24.3	4.9
	4021[a]	33.7	9.0

[a]From Taoukis and Labuza (1989a)
[b]From Wells and Singh (1988b)
[c]From Sherlock et al. (1991). Consumer-readable tags.
[d]Based on the data in McMeekin and Olley (1986)
[*]More models are available with similar kinetic parameters.

Thus, with the assumption of $T_{eff(food)} = T_{eff(TTI)}$, the effective rate constant and the microbial development for a known variable temperature exposure, $T(t)$, can be calculated through Eq. (10) and (1).

Similarly, one can introduce the concept of an equivalent time (t_{eq}), which is defined as the time at a reference temperature (k_{ref} as the rate constant) resulting in the same amount of quality changes as the variable time-temperature distribution. For food quality changes, $f(A)_t$, the following equation is obtained:

$$f(A)_t = \int_{t_1}^{t_2} k_A dt = k_{oA} \int_{t_1}^{t_2} \exp\left(-\frac{E_{A(food)}}{RT(t)}\right) dt = k_{ref(food)} t_{eq(food)} \qquad (11)$$

For a TTI exposed to the same time-temperature history and at the same reference temperature:

$$t_{eq(TTI)} = \frac{f(x)_t}{k_{ref(TTI)}} = \int_{t_1}^{t_2} \exp\left(-\frac{E_A(TTI)}{R}\left(\frac{1}{T(t)} - \frac{1}{T_{ref}}\right)\right) dt \qquad (12)$$

where $f(x)_t$ is the TTI response. Therefore the amount of food quality change or microbial development for a certain time-temperature history can be calculated from Eq. (11) based on the assumption that $t_{eq(food)} = t_{eq(TTI)}$ at the same reference temperature.

This equivalent time approach will be able to provide direct information on lost or remaining shelf-life of the food at a targeted storage temperature after exposure to a variable time-temperature history. For a continuously responding TTI, the difference between t_{eq} and t is the shelf-life loss due to the temperature abuse. The remaining shelf-life is equal to the open date minus t_{eq} if the open date is set based on the shelf-life at the targeted storage temperature. The value of t_{eq} is the result of the temperature abuse in the case of using a CTI tag, thus the remaining shelf-life is equal to the open date minus t_{eq} minus t, where t is the storage time.

For both the effective temperature approach and the equivalent time approach discussed above, there is an important underlying assumption, i.e., $T_{eff(food)} = T_{eff(TTI)}$ or $t_{eq(food)} = t_{eq(TTI)}$, respectively, for a given temperature distribution. This is true only when $E_{A(food)} = E_{A(TTI)}$ or when the temperature is constant throughout the distribution. The smaller the difference between $E_{A(food)}$ and $E_{A(TTI)}$, the more accurately the TTI indicates actual shelf-life (Taoukis and Labuza 1989). There are only a few TTI tags available today that have an activation energy value close to that for microbial growth.

To get around this problem, the equivalent point approach may be applied. Two or more TTI tags with different temperature sensitivities (i.e., different activation energy values) may be used simultaneously; they need not match that of the growth rate temperature dependence of the microbe being monitored. For the TTI response one can write:

$$f(x) = k_{ol}exp[-E_A(TTI)/(RT)]$$ (13)

Let $f(x)/k_{ol} = Y$, then the above equation can be transformed into:

$$\ln Y = -E_{A(TTI)}/(RT) + \ln t$$ (14)

A plot of $\ln Y$ vs. $E_{A(TTI)}$ gives a straight line. From the slope, T_e can be calculated and t_e can be calculated from the intercept. This shows that there is a unique point (T_e, t_e) for a variable time-temperature distribution. Thus, the level of microbial growth can be predicted for the same variable time-temperature distribution where:

$$N=N_oexp[-E_{A(food)}/(RT_e)(t_e-t_{Le})]$$ (15)

where t_{Le} is the lag time at the temperature of T_e.

The advantage of this approach is that it eliminates the strict requirement of $E_{A(food)} = E_{A(TTI)}$. However, it may be too costly to use two or more different types of TTI tags on the same case or package. One could design a multi-component tag which contains two or more chemical or biological components showing different temperature sensitivities (Maesmans *et al.* 1993) or to design a tag which will be able to match $E_{A(food)}$ by selecting the right reactive substrate (Fu *et al.* 1992). There are several companies now approaching both designs.

Temperature abuse may cause a history effect problem which could invalidate the results of a useful TTI. A history effect is one in which the previous time-temperature history has an effect on the microbial growth rate at the following storage condition, i.e., it grows faster or slower that predicted from the model for the next stage. Fu *et al.* (1991), showed that for microbial growth in milk there generally was a negative history effect, i.e., it took the organisms longer to attain the predicted growth rate. There is scant literature for fluctuating temperature conditions, but if the negative trend is generally valid, a TTI would underestimate actual shelf-life or time to an unsafe condition. This is more acceptable than overestimating the time.

Further information will also be required for TTIs including their reliability, accuracy, precision, sensitivity to temperature change, linearity under isothermal storage, integrating performance under variable time-temperature conditions, stability prior to activation, and ease of reading. The Campden Food and Drink Association in the UK has recently published the set of guidelines shown in Table 3 for the statistical evaluation of TTIs that appears be useful and could be incorporated into ISO 9001 standards (George and Shaw 1992). Additional TTI designs and varieties are expected to meet the requirements of different products with different activation energies (Boeriu *et al.* 1986; Fu *et al.* 1992). More complicated monitoring devices that may be able to monitor several parameters simultaneously may be developed in the near future (McMeekin *et al.* 1992).

TABLE 3.
TECHNICAL STANDARDS AND PROCEDURES FOR THE EVALUATION OF
TIME-TEMPERATURE INDICATORS A

Test procedure	Technical standard
Temperature response test	
Frozen food	Reproducible end point
Temperatures: -25, -15, -10, -5 and $+5°C$	Maximum tolerance: $\pm$ 6 days or 2.5% of the lifespan of the TTI
Chilled food	Reproducible end point
Temperatures: -5, $+5$, $+10$, $+15$ and $+25°C$	Maximum tolerance: $\pm$ 6 h or 2.5% or the lifespan of the TTI
Evaluation of kinetic parameters of TTI	
Make an Arrhenius plot based on the data collected in the above temperature response test	Arrhenius plot and the values of activation energy and preexponential factor
Temperature cycling test	
Frozen food	The predicted rate constant at the
Temperature range: $-25 \leftrightarrow 15°C$,	reference temperature from the
$-15 \leftrightarrow -10°C$, $-10 \leftrightarrow -5°C$, $-25 \leftrightarrow +5°C$	Arrhenius equation differs with
Cycling period: 10 times the intended lifespan of a TTI	the actually measured value by less than 10%.
Chilled food	
Temperature range: $-5 \leftrightarrow +5°C$,	Same as above
$+5 \leftrightarrow 10°C$, $+10 \leftrightarrow +15°C$, $-5 \leftrightarrow +25°C$	
Cycling period: Same as above	
Accuracy of initial activation	
Statistical quality control	Depends on batch size, sample size and the acceptable quality level (e.g., 1.5%)
Simulated field test	
To simulate actual use	Applicable and reliable
Other tests	TTI response should not be
Light (ultraviolet, visible and infrared)	affected by these factors
Vibration (including noise levels)	
Humidity	
Abuse tests (e.g., drop test)	

[a]From George and Shaw (1992)

Commercial Use of TTI

The application of TTI tags for monitoring the distribution chain and predicting the safe shelf-life of chilled and/or MAP meat and poultry also requires reliable microbial growth and shelf-life data. An estimate of the average initial quality is needed to do any prediction. The initial quality is determined by the presence of naturally occurring contaminating microorganisms, which may be very difficult to evaluate accurately and rapidly. In addition, different organisms have different temperature sensitivities and may compete with each other. Different assays are needed to differentiate temperature abused versus adequately refrigerated samples in a number of the products (Buchanan *et al.* 1992). More importantly, the current microbiological analytical techniques can require up to five days to identify a specific pathogen. This is too long for quality assurance testing of chilled MAP foods. Even using the impedance microbiological technique, determination of temperature abuse can take up to 24 h (Russell *et al.* 1992). Rapid microbial analysis techniques most likely employing the new biotechnologies are definitely needed (Giese 1993).

If one cannot detect the organisms directly, a TTI can be used to monitor the temperature exposure of meat and poultry products during distribution and up to the time they are displayed at the supermarket or even to the consumer's home and use that information to predict the potential for microbial presence. By attaching the monitor to individual packages or placing them in the immediate environment, they can give a measure of the effect of the preceding time-temperature history at each receiving place. The information gathered from all these places could be used for overall monitoring of the distribution channel, thus allowing for recognition and possible correction of the weak points. A TTI would allow targeting of responsibility and guarantee the producer and distributor that they can deliver a properly handled product to the retailer, and instruct the consumer in the importance of proper temperature storage at home or how soon the product should be used. Since meat frequently will be frozen at home after being purchased, a "freeze-by" date would be appropriate for open dating or "use-by" date for refrigeration, possibly coupled with a consumer tag.

There are at least three quality parameters that need to be evaluated in MAP products: microbial safety, microbial spoilage and overall organoleptic changes. However, the shelf-life of a product at any given temperature is determined by the mechanism which proceeds fastest and thus results in the shortest life. When safety becomes a major consideration in marketing the product, all efforts must be made to accurately determine its safe shelf-life. Figure 3 shows possible relationships using the log shelf-life versus temperature, between organoleptic spoilage (due to microbial spoilage, color change, etc.) and potential hazard (due to pathogenic microbial growth and toxin production). Case (a) is safe since spoilage occurs prior to pathogenic microbial growth at all possible storage

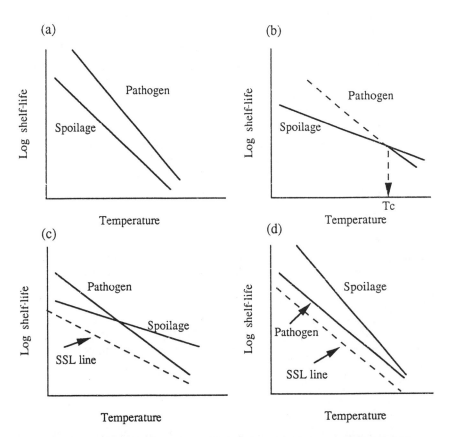

FIG. 3. POSSIBLE RELATIONSHIPS BETWEEN ORGANOLEPTIC SPOILAGE AND
POTENTIAL HAZARD DUE TO PATHOGEN GROWTH OR TOXIN PRODUCTION

temperatures. The kinetic data for spoilage can be used to incorporate an
appropriate TTI tag. Case (b) shows that the food is safe if stored below a
critical temperature, where no pathogen will grow or the growth of pathogenic
microorganisms to a hazardous level is slower than spoilage. A critical
temperature indicator could be used to serve as a warning sign that the T_c has
been exceeded for an unacceptable length of time. Case (c) shows the situation
where the outcome of spoilage or a microbial hazard depends on the temperature
of holding, i.e., above a critical temperature the pathogenic limit occurs before
the food is organoleptically spoiled. The dashed line is an assumed safe shelf-life
(SSL) line, which could be used for TTI correlation. As noted, it is drawn at a

shorter time for all temperatures. Case (d) indicates a product in which the pathogenic limit is reached before food spoilage, at all temperatures. Since setting the shelf-life at this limit could lead to a potential safety problem if the data have typical biological variability, an assumed SSL line parallel to the pathogen line could be used for TTI correlation. In cases (c) and (d), the key problem is how to determine the safety margin, i.e., the SSL line. Kraft General Foods, Inc., used a safety margin of between 1/3 to 1/4 of the product's organoleptic shelf-life as their printed shelf-life date (Harris 1989).

For fresh red meat the two most important quality properties controlling shelf-life are color and microbial population. The color of fresh meat depends on the relative amounts of three forms of myoglobin: reduced myoglobin, oxymyoglobin, and metmyoglobin (Young et al. 1988), which are controlled by the presence of O_2. TTIs generally cannot be used to monitor the color change of meat, since it is much more O_2 dependent than temperature dependent. Thus the packaging method and packaging material used are the main factors which control the amount of O_2 in the package headspace and thus the color of the meat. The microbial population of fresh meat is affected by many factors such as number and distribution of microbial species present at the start, health and handling of the live animal, slaughtering practices, chilling of the carcass, sanitation, type of packaging, and handling through distribution and storage (Young et al. 1988; Lambert et al. 1991b). During chilling of the carcasses, the relatively low temperatures give rise to predominately psychrotrophic bacteria such as Pseudomonas and Achromobacter. Lactobacillus organisms, which are facultative anaerobes, tend to thrive in vacuum-packaged meats. The dominant spoilage organisms differ depending on the storage temperature, the gas composition, and meat tissues (Jones 1989; Gill and Molin 1991).

Gill and Harrison (1989) studied the effects of vacuum and CO_2 on stability of pork cuts. Vacuum-packaged cuts were grossly spoiled by Brochothrix thermosphacta at both $3°C$ and $-1.5°C$, as were cuts packaged under CO_2 at $3°C$. Growth of B. thermosphacta was suppressed in CO_2 packaged cuts at $-1.5°C$. At that temperature, the enterobacteria caused gross spoilage of the cuts. Figure 4 is the shelf-life plot for pork, from which the activation energy and Q_{10} are calculated and listed in Table 4. These data can be used in choosing an appropriate tag. This situation is a good example of case (a) from Fig. 3. Based on the data in Table 2, there are essentially only one or two tags with the same high temperature sensitivity.

Clark and Lentz (1969) reported that the growth of psychrotolerant, slime-producing bacteria (Pseudomonas and Achromobacter) on the surface of fresh beef was inhibited by CO_2. The degree of inhibition depended on the temperature, CO_2 concentration and the age of the culture at the time of CO_2 application. It was found that 20% CO_2 was the most practical concentration, since it was effective at temperatures as high as $10°C$, and since higher concentrations

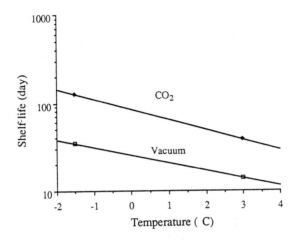

FIG. 4. SHELF-LIFE PLOT FOR MAP PORK CUTS BASED ON THE RESULTS
OF GILL AND HARRISON (1989)

TABLE 4.
ACTIVATION ENERGY AND Q_{10} VALUES OF VARIOUS MAP FOOD PRODUCTS

Food system	MA condition	E_A (kcal/mole)	Q_{10}
Raw pork[a]	vacuum	31.3	7.7
	CO2	40.4	13.9
Raw beef[b]	air	15.8	2.8
	10% CO2	20.3	3.7
	20% CO2	19.2	3.5
	30% C°2	19.9	3.6
Raw chicken[c]	air	15.9	2.8
	30% CO2 + air	23.5	4.6
	30% CO2 + N2	23.5	4.6
	100% CO2	27.1	5.8

[a]Based on data of Gill and Harrison (1989)
[b]Based on the data of Clark and Lentz (1969)
[c]Based on the data of Hart et al. (1991)

caused color change in the meat but gave little additional inhibition of growth. Figure 5 is the shelf-life plot for fresh beef packed under modified atmospheres. As seen, the level of CO_2 does not change the temperature sensitivity very much, rather only affecting out-growth time. The related kinetic parameters are listed in Table 3. Since the CO_2 environment in a package can change with time, 20% CO_2 can be used to choose a tag. This situation is best represented by case (a) of Fig. 3 in which spoilage occurs before outgrowth of pathogens. Based on Table 2, there are several tags with the same or similar temperature sensitivity.

Hart *et al.* (1991) studied the effects of gaseous environment and temperature on the storage behavior of *Listeria monocytogenes* in chicken breast meat. The results of the study suggested that contamination with *Listeria monocytogenes* would not lead to any significant growth prior to spoilage under normal conditions of handling and storage. Holding the chilled meat under 100% CO_2 was the most effective means of extending the shelf-life while ensuring that levels of *L. monocytogenes* were also suitably controlled. Figure 6 is the shelf-life plot for chicken based on "off" odor development. The data for 100% CO_2 in Table 3 should be used for TTI correlation. This is similar to case (a) or (b) of Fig. 3.

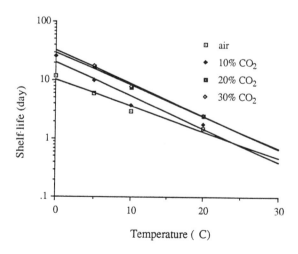

FIG. 5. SHELF-LIFE PLOT FOR MAP FRESH BEEF BASED ON THE
RESULTS OF CLARK AND LENTZ (1969)

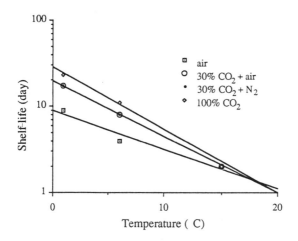

FIG. 6. SHELF-LIFE PLOT FOR MAP FRESH CHICKEN BASED ON THE
RESULTS OF HART *ET AL.* (1991)

In a conflicting study, Wimpfheimer *et al.* (1990) showed that MAP of raw
chicken can substantially inhibit the aerobic spoilage microorganisms while
allowing the levels of pathogenic *L. monocytogenes* to increase. Fortunately, the
organism failed to grow at all on meat stored at 1–6°C under 100% CO_2, which
is relevant to UK commercial practice for bulk storage of poultry meat under
CO_2. But the beneficial effects of CO_2 were lost when the meat was exposed to
an overt abuse temperature of 15°C. This temperature may be used as the
critical temperature for choosing a critical temperature indicator, as in case (b)
of Fig. 3. Thus a dual tag would be needed to indicate organoleptic shelf-life
and potential pathogen presence.

Based on the data reported by Wimpfheimer *et al.* (1990), the growth of
aerobic microorganisms and *L. monocytogenes* on raw chicken under air had E_A
values of 32.4 and 20.0 Kcal/mol, and under MA (72.5:22.5:5, CO_2:N_2:O_2) had
values of 23.2 and 21.0 Kcal/mol, respectively. The data showed that MA did
not affect the temperature sensitivity of *L. monocytogenes* while influencing the
aerobic microorganisms significantly. When the temperature is lowered, the
aerobic microorganisms are inhibited and *L. monocytogenes* may grow freely;
thus a hazard issue arises. At this point, optimization of MA conditions is
necessary. For example, use of 100% CO_2 can avoid this problem, and a critical
temperature tag, set to 14°C (below 15°C) to be safe, would be valuable.

Zamora and Zaritzky (1985) presented mathematical equations to estimate the time necessary for refrigerated vacuum packaged beef to reach a bacterial density of 10^7 cfu/cm^2 under various storage conditions. Initial composition of the mixed flora, individual growth rates and lag times were obtained experimentally. The growth rate and lag time for each microorganism were functions of temperature, pH of the meat cut and permeability of the packaging film. Their model permitted a satisfactory prediction of microbial growth for any constant, but did not include changing the storage temperature during storage.

In another study, commercially prepared sliced roast beef in conventional vacuum and oxygen-free saturated carbon dioxide controlled atmosphere packs was stored at 10, 3 and $-1.5\,°C$ (Penney et al. 1993). The development of foreign flavors during chilled storage resulted in product rejection before the development of putrid flavors characteristic of overt microbial spoilage. Product in vacuum packs was rejected on sensory criteria by a trained panel after 3 days, 3 weeks and 8 weeks for storage at 10, 3 and $-1.5\,°C$, respectively. At these temperatures, in oxygen-free saturated carbon dioxide packages, rejection levels were not attained after 4 days, 10 weeks and 16 weeks, respectively. These shelf-life data essentially determine what type of TTIs should be used.

More shelf-life data for meat and poultry products are available in the shelf-life dating book by Labuza (1982). Unfortunately, little information is available on the competitive growth of pathogen and spoilage organism in real food products. There is a need not only for additional research on the storage life, but also on the safety of these products up to and beyond the manufacturer's dated storage life at different temperatures, so that the use of a TTI can be evaluated.

Electronic Temperature Recorders or Monitors

For many years, the food industry has temperature recording devices for monitoring distribution temperatures, especially for frozen or refrigerated foods (Anon. 1992). Electronic devices which record temperature and are programmed with the appropriate microbial response offer an alternative. They record temperature and then analyze that temperature history with specific computer software to predict the extent of microbial growth at any time, or over any time interval in that products' history. Thus elapsed shelf-life, or pathogen growth, or the effects of a particular part of that history on product quality can be estimated using one of the above models. Unfortunately some manufacturers use the area under the T-t curve as the model which we have shown to be incorrect, and a few commercial loggers and integrators incorporate square root type relationships in their programming (McMeekin et al. 1992). The main reason for limited use is that all of these recorder devices are relatively expensive ($20 to > $500) and monitor the environment of a case, pallet load, truck, train, etc., and not necessarily the individual food pack which would be required for information aimed at the consumer.

Combined relative humidity and temperature loggers are commercially available (McMeekin *et al.* 1992). There is also potential for the development of combined sensors to record temperature and other significant variables. More complicated computer program is needed to interpret these combined effects.

RESEARCH NEEDS

A consumer TTI tag plus an open date code will give more information to the consumer about the stability or safety of a meat or poultry product. Use of TTIs will influence buying patterns and will help to develop consumer confidence in the meat and poultry industry. The use of TTIs can help the overall industry achieve a safer and higher quality product. However, the successful application of a TTI tag or a package biosensor depends on many factors, such as product stability analysis, reliability of a TTI, careful handling by the distributor and retailer and cooperation of consumers, major breakthroughs in thin film biotechnology and knowledge of microbial surface logistics. Improvements in the preparation and handling of meat and poultry products monitored by HACCP could also result in significant improvement in shelf-life and reductions in foodborne illness. Studies are needed to develop an effective HACCP plan for a meat/poultry plant and for a food service unit. This in combination with on-line or on-package sensors or a TTI should lead to a safer and wholesome meat and poultry supply (Giese 1993).

ACKNOWLEDGMENTS

The research was supported in part by the Minnesota/South Dakota Dairy Research Center. It is published as a paper of the contribution series of the Minnesota Agricultural Experiment Station.

REFERENCES

AFDO. 1991. Refrigerated foods in reduced oxygen packages. Assoc. of Food and Drug Official Retail Guidelines, pp. 80–84.

ALMONACID-MERINO, S.F. and TORRES, J.A. 1993. Mathematical models to evaluate temperature abuse effects during distribution of refrigerated solid foods. J. Food Eng. *20*, 223–245.

ANON. 1989. Is it time for time/temperature indicators? Prep. Foods *158*(11), 219.

ANON. 1992. Transit Temperature Recording—Origin and History. Long Beach, CA: Cox Recording Co.

ANON. 1993. Mandatory safe handling statement on labeling of raw meat and poultry products. Fed. Reg. *58* (156), 9 CFR Parts 317 and 381.

BAKER, D.A. and GENIGEORGIS, C. 1990. Predicting the safe storage of fresh fish under modified atmospheres with respect to *Clostridium botulinum* toxigenesis by modeling length of the lag phase of growth. J. Food Prot. *53*(2), 131–140.

BELL, L.N., FU, B. and LABUZA, T.P. 1992. Criteria for experimental design and prediction of food shelf-life. In *Advances in Food Engineering*, (R.P. Singh and M. Wirakartakusumah, eds.), CRC Press, London.

BLIXT, K.G. 1983. The I-point® TTM—A versatile biochemical time-temperature integrator. In *Proceedings 16th Intl. Cong. Refrig.*, pp. 629–631, IIR Commission C2 Preprints.

BOERIU, C.G., DORDICK, J.S. and KLIBANOV, A.M. 1986. Enzymatic reactions in liquid and solid paraffins: application for enzyme-based temperature abuse sensors. Bio/Technol. *4*, 997, 999.

BRATCHELL, N., GIBSON, A.M., TRUMAN, M., KELLY, T.M. and ROBERTS, T.A. 1989. Predicting microbial growth: The consequences of quantity of data. Int. J. Food Microbiol. *8*, 47–58.

BUCHANAN, R.L. 1993. Predictive food microbiology. Trends Food Sci. Technol. *4*(1), 6–11.

BUCHANAN, R.L. and PHILLIPS, J.C. 1990. Response surface model for predicting the effects of temperature, pH, sodium chloride content, sodium nitrite concentration, and atmosphere on the growth of *Listeria monocytogenes*. J. Food Prot. *53*, 370–376.

BUCHANAN, R.L., SHULTZ, F.L., GOLDEN, M.H., BAGI, K. and MARMER, B. 1992. Feasibility of using microbiological indicator assays to detect temperature abuse in refrigerated meat, poultry, and seafood products. Food Microbiol. *9*, 279–301.

CLARK, D.S. and LENTZ, C.P. 1969. The effect of carbon dioxide on the growth of slime producing bacteria on fresh beef. Can. Inst. Food Technol. J. *2*(2), 72–75.

CORLETT, D.A. Jr. 1989. Refrigerated foods and use of hazard analysis and critical control point principles. Food Technol. *43*(2), 91–94.

DOYLE, M.P. 1991. Evaluating the potential risk from extended-shelf-life refrigerated foods by *Clostridium botulinum* inoculation studies. Food Technol. *45* (4), 154–156.

FIELDS, S.C. and PRUSIK, T. 1986. Shelf life estimation of beverage and food products using bar coded time-temperature indicator labels. In *The Shelf Life of Foods and Beverages*, (G. Charalambous, ed.) pp. 85–96, Elsevier Science Publishers, Amsterdam.

FU, B. and LABUZA, T.P. 1992. Considerations for the application of time-temperature integrators in food distribution. J. Food Distr. Res. *23*(1), 9–17.

FU, B. and LABUZA, T.P. 1993. Shelf-life prediction: theory and application. Food Control *4*(3), 125–133.

FU, B., TAOUKIS, P.S. and LABUZA, T.P. 1991. Predictive microbiology for monitoring spoilage of dairy products with time-temperature integrators. J. Food Sci. *56*, 1209–1215.

FU, B., TAOUKIS, P.S. and LABUZA, T.P. 1992. Theoretical design of a variable activation energy time-temperature integrator for prediction of food or drug shelf life. Drug Dev. Indu. Pharm. *18*, 829–850.

GAUCH, JR. H.G. 1993. Prediction, parsimony and noise. Am. Scientist *81*, 468–478.

GEORGE, R.M. and SHAW, R. 1992. A food industry specification for defining the technical standard and procedures for the evaluation of temperature and time temperature indicators. Tech. Manual No. 35. Campden Food & Drink Research Assoc., London.

GIBSON, A.M., BRATCHELL, N. and ROBERTS, T.A. 1987. The effect of sodium chloride and temperature on the rate and extent of growth of *Clostridium botulinum* type A in pasteurized pork slurry. J. Appl. Bacteriol. *62*, 479–490.

GIESE, J. 1993. Rapid techniques in quality assurance. Food Technol. *47*(10) 52–60.

GILL, C.O. and HARRISON, J.C.L. 1989. The storage life of chilled pork packaged under carbon dioxide. Meat Sci. *26*(4), 313–324.

GILL, C.D. and MOLIN, G. 1991. Modified atmosphere and vacuum packaging. In *Food Preservatives* (N.J. Russell and G.W. Gould, eds.) pp. 172–179. Blackie and Son, Glasgow.

GOMBAS, D.E. 1989. Biological competition as a preserving mechanism. J. Food Safety. *10*(2), 107–117.

HARRIS, R.D. 1989. Kraft builds safety into next generation refrigerated foods. Food Proc. *50*(12), 111–114.

HART, C.D., MEAD, G.C. and NORRIS, A.P. 1991. Effects of gaseous environment and temperature on the storage behavior of *Listeria monocytogenes* on chicken breast meat. J. Appl. Bacteriol. *70*(1), 42–46.

HINTLIAN, C.B. and HOTCHKISS, J.H. 1986. The safety of modified atmosphere packaging: A review. Food Technol. *40*(4), 70–76.

JONES, M.V. 1989. Modified atmospheres In *Mechanisms of Action of Food Preservation Procedures*, (G.W. Gould, ed.) pp. 247–284. Elsevier Applied Science, London.

LABUZA, T.P. 1982. *Shelf-Life Dating of Foods*, Food & Nutrition Press, Trumbull, CT.

LABUZA, T.P. 1984. Application of chemical kinetics to deterioration of foods. J. Chem. Edu. *61*, 348–358.

LABUZA, T.P. 1993. Glowing future for poultry — after the change. Poultry Proc. April/May 22–24, 26, 28, 30, 32.

LABUZA, T.P. 1994. Active packaging technologies for improvement of quality and shelf life. In *Science for the Culinary Industry of the 21st Century* (M. Yalpani, ed.) ATL Press Sci. Pub., Chicago.

LABUZA, T.P., FU, B. and TAOUKIS, P.S. 1992. Prediction for shelf life and safety of minimally processed CAP/MAP chilled foods. J. Food Prot. *55*, 741–750.

LAMBERT, A.D., SMITH, J.P. and DODDS, K.L. 1991a. Effect of headspace CO_2 concentration on toxin production by *Clostridium botulinum* in MAP, irradiated fresh pork. J. Food Prot. *54*, 588–592.

LAMBERT, A.D., SMITH, J.P. and DODDS, K.L. 1991b. Shelf life extension and microbiological safety of fresh meat: a review. Food Microbiol. *8*, 267–297.

LEE, Y.C., SINGH, R.P. and HAARD, N.F. 1992. Changes in freshness of Chillipeper Rockfish during storage as measured by chemical sensors and biosensors. J. Food Biochem. *16*, 119–129.

MAESMANS, G., HENDRICKX, M., DE CORDT, S. and TOBBACK, P. 1993. Theoretical considerations on design of multicomponent time temperature integrators in evaluation of thermal processes. J. Food Processing Preservation. *17*, 369–389.

MANSKE, W.J. 1983. The application of controlled fluid migration to temperature limit & time temperature integrators. In *Proceedings of 16 Intl. Cong. Refrig.* pp. 632–635, IIR Commission C2 Preprints.

McMEEKIN, T.A., ROSS, T. and OLLEY, J. 1992. Application of predictive microbiology to assure the quality and safety of fish and fish products. Int. J. Food Microbiol. *15*, 13–32.

MONOD, J. 1949. The growth of bacterial cultures. Ann. Rev. Microbiol. *3*, 371–394.

NOTERMANS, S., DUFRENNE, J. and LUND, B.M. 1990. Botulism risk of refrigerated, processed foods of extended durability. J. Food Prot. *53*, 1020–1024.

OGILVY, W.S. and AYRES, J.C. 1951. Post-mortem changes in stored meats. II. The effect of atmospheres containing carbon dioxide in prolonging the storage life of cut-up chicken. Food Technol. *5*(3), 97–102.

PENNEY, N., HAGYARD, C.J. and BELL, R.G. 1993. Extension of shelf-life of chilled sliced roast beef by carbon dioxide packaging. Int. J. Food Sci. Technol. *28*, 181–191.

POONI, G.S. and MEAD, G.C. 1984. Prospective use of temperature function integration for predicting the shelf life on non-frozen poultry-meat products. Food Microbiol. *1*, 67–78.

RATKOWSKY, D.A., LOWRY, R.K., McMEEKIN, T.A., STOKES, A.N. and CHANDLER, R.E. 1983. Model for bacterial culture growth rate throughout the entire biokinetic temperature range. J. Bacteriol. *154*, 1222–1226.

RATKOWSKY, D.A., OLLEY, J., McMEEKIN, T.A. and BALL, A. 1982. Relationship between temperature and growth rate of bacterial cultures. J. Bacteriol. *149*(1), 1–5.

RUSSELL, S.M., FLETCHER, D.L. and COX, N.A. 1992. A rapid method for the determination of temperature abuse of fresh broiler chicken. Poultry Sci. *71*, 1391–1395.

SHERLOCK, M., FU, B., TAOUKIS, P.S. and LABUZA, T.P. 1991. A systematic evaluation of time-temperature indicators for use as consumer tags. J. Food Prot. *54*, 885–889.

TAOUKIS, P.S., FU, B. and LABUZA, T.P. 1991. Time–temperature indicators. Food Technol. *45*(10), 70–82.

TAOUKIS, P.S. and LABUZA, T.P. 1989. Applicability of time-temperature indicators as food quality monitors under non-isothermal conditions. J. Food Sci. *54*, 783–788.

TOMPKIN, R.B. 1990. The use of HACCP in the production of meat and poultry products. J. Food Prot. *53*, 795–803.

WELLS, J.H. and SINGH, R.P. 1988a. Application of time-temperature indicators in monitoring changes in quality attributes of perishable and semiperishable foods. J. Food Sci. *53*, 148–152.

WELLS, J.H. and SINGH, R.P. 1988b. Response characteristic of full-history time-temperature indicators suitable for perishable food handling. J. Food Processing Preservation *12*, 207–218.

WILLCOX, F., MERCIER, M., HENDRICKX, M. and TOBBACK, P. 1993. Modeling the influence of temperature and carbon dioxide upon the growth of *Pseudomonas fluorescens*. Food Microbiol. *10*, 159–173.

WIMPFHEIMER, L., ALTMAN, N.S. and HOTCHKISS, J.H. 1990. Growth of *Listeria monocytogenes* Scott A, Serotype 4 and competitive spoilage organisms in raw chicken packaged under modified atmospheres and in air. Int. J. Food Microbiol. *11*, 205–214.

YOUNG, L.L., REVIERE, R.D. and COLE, A.B. 1988. Fresh red meats: a place to apply modified atmospheres. Food Technol. *42*(9), 65–69.

ZAMORA, M.C. and ZARITZKY, N.E. 1985. Modeling of microbial growth in refrigerated packaged beef. J. Food Sci. *50*, 1003–1013.

ZWIETERING, M.H., De KOOS, J., HASENACK, B.E., De WIT, J.C. and van't RIET, K. 1991. Modeling of bacterial growth as a function of temperature. Appl. Environ. Microbiol. *57*, 1094–1101.

ZWIETERING, M.H., JONGENBURGER, I., ROMBOUTS, F.M. and van't RIET, K. 1990. Modeling of the bacterial growth curve. Appl. Environ. Microbiol. *56*, 1875–1881.

CHAPTER 12

RELATIONSHIPS BETWEEN PATHOGEN GROWTH AND THE GENERAL MICROBIOTA ON RAW AND PROCESSED MEAT AND POULTRY

MARY UPTON

Department of Industrial Microbiology
University College, Dublin
Belfield, Dublin 4, Ireland

ABSTRACT

Contaminants are capable of growth on meat and poultry products due to the nutrient composition of the products, the available water, the favorable pH and very often the temperature and other conditions of storage. These factors and processing conditions influence greatly the pattern of development of the spoilage and pathogenic microflora. This paper will consider the effects of the general bacterial flora on pathogenic bacteria occurring on meat and poultry as well as the impact of external factors on some of the different types of bacteria found on these products.

FACTORS AFFECTING MICROBIAL GROWTH

Microbial growth patterns on foods are affected by intrinsic, extrinsic, processing and implicit factors (Mossel 1983). Intrinsic factors include the concentration and availability of nutrients, pH and buffering capacity, redox potential and water availability (Aw). Extrinsic factors are concerned mostly with storage conditions and include storage temperature and the composition and relative humidity of the gaseous environment surrounding the food. The rate at which the food equilibrates with the extrinsic factors, for example as a result of temperature change or atmosphere modification, will be governed by its intrinsic structural properties (Brown 1982). When food is not at equilibrium with its surroundings, gradients of gas concentration, Aw and temperature influence types of bacteria growing at and beneath the surface.

Most processing procedures lead to the elimination of microorganisms or a reduction in numbers. Heat processing is a typical example of this effect. Fermentation which will not necessarily eliminate pathogens from a product but it may help to reduce or control the numbers.

Implicit factors in food spoilage include microbial antagonism or synergism. Microbial antagonism is much more significant than synergism. Antagonism may occur when one group or population of microorganisms produces a substance inhibitory to other species or groups. Bacteriocin production, for example, may have this effect.

Competition may occur where both populations are restricted because of their common dependence on a limiting factor, such as a nutrient. Antagonism and competition are probably the most common interactions and often result in the dominance of one species. Boddy and Tempanny (1992) point out, however, that frequently interaction between two bacterial species is considered experimentally, but in reality interactions usually occur between numerous species/individuals. This situation receives little attention. Further consideration should be given to the interactions between populations on such complex environments as meat or poultry, where a large number of different species are initially present together on the product.

MICROBIOLOGY OF MEAT AND POULTRY

The condition of the animal at slaughter, the spread of contamination during slaughter and processing, and the temperature, time and other conditions of storage and distribution are the main determinants of meat and poultry quality. The initial microbiota usually consists of a wide range of organisms adapted to whatever conditions prevail as a result of processing and storage. The microbiota that predominates is mainly a reflection of the environment.

Microbial Pathogens

Meat and poultry are frequently implicated in the spread of food-borne disease. *Salmonella* is one of the major concerns and *Salmonellae* from poultry and meat products is common in many countries. Salmonella infection is often spread among animals or poultry through the use of contaminated feed and the incidence tends to be highest where intensive stock raising is practiced. Problems with *Salmonella* can be exacerbated by stress and starvation during transport. Hygiene during slaughter and dressing together with prompt and adequate cooling are important (Brown 1982; Cooper 1994).

Food poisoning Staphylococci are widely distributed and meat and poultry can become contaminated from human as well as animal or poultry sources (Bryan 1968). Other pathogens which may be present include *Listeria monocytogenes* (Johnson *et al.* 1990), *Campylobacter jejuni* (Harris *et al.* 1986), *Yersinia enterocolitica* (Hanna *et al.* 1976), *Clostridium perfringens* (Lillard 1971) and *E. coli 0157:H7* (Doyle and Schoeni 1989).

Spoilage Microbiota

In addition to the microorganisms present in the tissues of symptomless carriers there are many other contamination sources which contribute to the microbial load on meat and poultry. The skin or hide of the particular animal or poultry and others being dressed nearby, is probably the major source of contamination. Much of this contamination is originally of fecal origin but it will include the normal flora of the skin — staphylococci, micrococci, pseudomonads, yeasts and molds as well as organisms from soil and water. Many of these organisms are psychrotrophs, able to grow at low temperatures and are potential spoilers of chilled meat and poultry. The bacterial biota of chilled meats and poultry includes *Pseudomonas spp., Moraxella spp.*, Enterobacteriaceae, and the gram positive *Brochothrix thermosphacta* (Dainty *et al.* 1983).

The extent to which contamination occurs and the composition of the resulting biota flora reflects the hygiene standard in the slaughterhouse and the cleanliness of the animal. In addition to skin, the gastrointestinal and respiratory tracts, knives, cloths, hands and clothing of workers, processing equipment and the water used to wash carcasses also contribute (Bryan *et al.* 1980).

MODIFIED ATMOSPHERE PACKAGED MEAT AND POULTRY

The emergence of psychrotrophic pathogens such as *Listeria monocytogenes, Yersinia enterocolitica* and nonproteolytic strains of *Clostridium botulinum* raise significant safety issues in relation to Modified Atmosphere Packaged (MAP) meat and poultry products. The extended shelf-life of many MAP products may allow extra time for pathogens to reach dangerous levels in a food.

L. monocytogenes has caused great concern to the food industry and to regulatory agencies. *L. monocytogenes* is a gram positive psychrotrophic pathogen capable of growth under MAP conditions. Johnson *et al.* (1990) reviewed the incidence of *L. monocytogenes* in meat and meat products and noted that it is a common isolate from such products. Tran *et al.* (1990) studied the effect of naturally occurring aerobic mesophilic microbiota concentration in a variety of foods (including some meats) on *Listeria* isolation. They concluded that when competition occurs specific bacterial competitors may be more important rather than bacterial numbers.

Marshall *et al.* (1991) compared growth of *L. monocytogenes* and *Pseudomonas fluorescens* on pre-cooked chicken nuggets under MAP conditions and showed that although *L. monocytogenes* was inhibited by MAP conditions it was still capable of growth. Growth of *Ps. fluorescens* was inhibited to a greater extent than *L. monocytogenes*.

Grau and Vanderlinde (1990) examined the growth pattern of *L. monocytogenes* in vacuum packaged beef at 0 and 5°C under different pH conditions and on lean and fatty tissue. In all treatments the normal psychrotrophic biota associated with vacuum packaged beef grew at a faster rate than did *Listeria*. However, *Listeria* was able to grow even when significantly outnumbered. For example, after four weeks storage, *Listeria* grew in purge fluid in the presence of more than 10^7 other bacteria per ml.

In 1992 Grau and Vanderlinde determined the effect of storage temperature on the growth rate of *L. monocytogenes* inoculated on to vacuum packed sliced corned beef and ham. As the storage temperature increased from 0 to 15°C, the growth rate of *L. monocytogenes* increased more rapidly than that of the other biota-lactic acid bacteria and *Brochothrix thermosphacta*. In addition to pH, salt and residual nitrite may influence the rate of growth of *L. monocytogenes* on chilled meats. Products of high pH, high Aw, and low residual nitrite are most likely to permit significant increases in Listeriae. Wimpfheimer *et al.* (1990) studied the growth of *L. monocytogenes* and competitive spoilage organisms in raw chicken packaged under modified atmospheres and in air. They showed that modified atmosphere packaging of raw chicken and probably other meats can substantially inhibit the aerobic spoilage biota while allowing pathogenic *L. monocytogenes* to grow.

Campylobacter jejuni is widely recognized as pathogenic for humans and it is frequently isolated from poultry. It is a normal commensal of all bovine classes and is a common contaminant of swine carcasses and sheep (Franco 1988). Thus its isolation from meat and poultry products is not surprising. Acute enteritis is a feature of *C. jejuni* infection with diarrhea, malaise, fever and abdominal pain as common symptoms (Blaser *et al.* 1984). *C. jejuni* grows best in an atmosphere containing 5-10% O_2 with about 8-10% CO_2 and 85% N_2 as the most favorable for growth. *C. jejuni* dies off rapidly at ambient temperatures and atmospheres, grows poorly on food and is not considered to be a good competitor because it does not utilize carbohydrates (Doyle and Jones 1992). However, it has been suggested ingestion of only a small number, ca.500 *C. jejuni* cells, may produce human illness (Black *et al.* 1988; Robinson 1981) and this small infectious dose may compensate for the organisms otherwise fragile and fastidious nature.

Because of its microaerophilic requirements *C. jejuni* could be expected to survive in certain MAP products. Phebus *et al.* (1991) inoculated turkey roll with *C. jejuni* and monitored its survival at 4 and 21°C under various atmospheric mixtures of N_2, CO_2 and O_2. They also studied the effect of different gas atmospheres on the growth rate and type of spoilage microbiota in turkey roll, especially psychrotrophs and lactic acid bacteria. They showed that a potential risk may exist with storage of turkey roll contaminated with *C. jejuni* in modified atmospheres containing high CO_2 levels. The atmospheres that were most protective to *C. jejuni* were also the most inhibitory to the natural spoilage organisms of the turkey roll. In a similar experiment Reynolds and Draughon (1987) showed that culturable *C. jejuni* decreased significantly during vacuum packaged storage on turkey and ham and turkey roll samples at 4°C and that aerobic plate counts and Enterococci increased significantly during storage, providing competition for *C. jejuni*. Although survival of *C. jejuni* decreased over time, greater than 500 viable cells per gram were detected with some strains for up to 28 days.

Yersinia enterocolitica is a major cause of enteric infection (WHO 1980). Ibrahim and MacRae (1991) described the isolation of *Y. enterocolitica* from beef, lamb and pork and Norberg (1981) recorded its isolation from frozen chicken. *Y. enterocolitica* is a facultatively anaerobic psychrotrophic rod belonging to the family *Enterobacteriaceae*. Swine seems to be the principal reservoir of virulent strains. *Y. enterocolitica* can grow on high pH beef packaged under 100% CO_2 at 5 and 10°C. Growth however does not occur at 0, 2 or −2°C (Gill and Reichel 1989). Kleinen and Untermann (1990) showed that at 15°C *Y. enterocolitica* growth was equal to that of the air control and only slightly delayed at 10°C but that the presence of a large competitive background biota inhibited the growth of *Y. enterocolitica* at all temperatures tested i.e., 1, 4, 10, and 15°C.

Most of the Clostridia that occur in raw meats are harmless putrefactive mesophiles. However, *Clostridium botulinum* spores are occasionally present and must be considered in the MAP storage of meat and poultry products (Lucke and Roberts 1992). Botulism from pork products is more frequent than from other meats. Poultry and poultry products have not been associated with human botulism. Botulism from meat is mainly related to inadequate home preserving, combined with poor storage and insufficient reheating. Lucke and Roberts comment that many detailed investigations have yielded inconclusive results as to whether toxin production by *Clostridium botulinum* can occur before spoilage. However, there are sufficient examples of toxigenesis prior to spoilage being evident, to state firmly that spoilage should not be relied upon before toxin production to prevent consumption. Many of these studies refer to fish products (Post *et al.* 1985; Steir *et al.* 1981) where there is clearly a greater risk of contamination from *C. botulinum*. The development of REPFEDS (Refrigerated Processed Foods with Extended Durability; Mossel and Thomas 1988) combined with MAP certainly requires further studies on *C. botulinum* on meat and poultry products.

Facultative anaerobic food pathogens such as *Staphylococcus aureus* and *Salmonella* spp. grow very slowly, if at all, at refrigeration temperatures. Modified atmospheres at refrigeration temperatures do not encourage the growth of these organisms (Luiten *et al.* 1982). The concern about these pathogens would therefore relate primarily to abuse conditions of storage (Hintlian and

Hotchkiss 1987). Modified atmospheres inhibit growth of *S. aureus* during abusive storage but are less effective against *Salmonella typhimurium*.

E. *coli 0157:H7* is a pathogen of relatively recent concern, which has been isolated from ground beef, pork, chicken, turkey and lamb (Doyle and Schoeni 1989). Several outbreaks of hemorrhagic colitis have been caused by this bacterium, the confirmed or suspected vehicle in most cases being ground beef. There is contradictory evidence regarding the ability of *E. coli 0157:H7* to grow at low temperatures. Doyle and Schoeni (1984) reported that the organism could not grow at 4°C whereas Lechowich (1988) reported that it grew at temperatures between 1.1 and 4.4°C. The ability of this bacterium to grow at low temperatures and under MAP conditions is of considerable significance and its behaviour and survival in the presence of a normal spoilage and competitive flora on meat and poultry should be studied in greater detail.

FERMENTED MEATS

The manufacture of fermented meats is characterized by microbial activity taking place within a solid/semi-solid matrix, with spatial distribution of microorganisms, mixing and heterogeneity of the resulting microenvironment, and mass transfer limitations on both solute and gas transfer, all playing a role. More significantly, the substrate is not usually sterilized or pasteurized, so the desired lactic acid bacteria (LAB) must compete with the indigenous microflora (McLoughlin and Champagne, in press).

While traditionally the production of lactic acid and the consequent drop in pH was considered to be the most significant factor in the preservation of fermented meats, more recently it is recognized that other products of the LAB, notably bacteriocins, and the production of H_2O_2, to a lesser extent, may contribute to the overall antibiosis and preservative potential (Lindgren and Dobrogosz 1990).

Several reports refer to the behavior of *L. monocytogenes* in fermented meats. Buncic *et al.* (1991) showed that numbers of *L. monocytogenes* declined most rapidly during the first four days of ripening, parallel with a drop in pH to 4.8. Johnson *et al.* (1988) examined the behavior of *L. monocytogenes* in hard salami made from inoculated meat. Numbers of *Listeria* decreased by ca.1 $\log_{10}$ cfu/g during fermentation, with another slight decrease occurring during the nine day drying period (Glass and Doyle 1989). They concluded that sausage fermentation and the drying process reduce but do not eliminate *Listeriae* from sausage. Schillinger and Lucke (1989) showed that *Lactobacillus sake Lb 706* excretes a bacteriocin, sakacin A, which is effective against *L. monocytogenes*. Schillinger, Kaya and Lucke (1991) studied the production of this bacteriocin in meat products, such as pasteurized minced pork, and in broth and showed inhibition in both environments but to a lesser extent in the meat than in the broth. In raw pork filled into casings, *L. monocytogenes* grew at pH 6.3 but addition of *L. sake Lb706* prevented the growth of Listeriae during the first few days after manufacture.

Goepfert and Chung (1970) studied the growth of *S. typhimurium* in simulated Thuringer sausage and found that although low pH and salt was effective in reducing numbers, it did not always produce a Salmonella free product. Baran and Stevenson (1975) found that at initial levels of 10^6/g *S. pullorum* and *S. senftenberg* were not eliminated during sausage fermentation or during subsequent processing. Smith and Palumbo (1981) concluded that as low numbers of Salmonella were often detected at the end of processing the LAB alone cannot always produce a Salmonellae-free fermented sausage.

Smith and Palumbo (1981) cite a number of references to food poisoning caused by dry fermented sausage where *S. aureus* was implicated. Daly *et al.* (1973) showed that commercial starter cultures do not entirely suppress growth of *S. aureus* but reduce growth markedly, and Metaxopoulos *et al.* (1981a and 1981b) noted that *S. aureus* numbers can increase significantly before LAB can bring about the required pH drop to prevent growth. Lewus *et al.* (1991) report that several strains of LAB isolated from meat produce bacteriocins active against *S. aureus* and suggest that bacteriocin-producing strains of LAB in meat products may suppress its growth.

In 1955 Saleh and Ordal attempted to control growth and toxin production of *Clostridium botulinum* using a mixture of strains of *Streptococcus* and *Leuconostoc* or *Lactobacillus bulgaricus* in sterilized Chicken a la King. Product tested after five days was toxic. Riemann and Genigeoris (1972) recommended that 1% glucose be added to semi-preserved meats, like ham or bacon, to reduce pH below a level which would support pathogen growth but they also suggested adding a lactic culture to ferment the glucose.

There is little literature about *E. coli 0157:H7* in fermented meats but Enteropathogenic *E. coli* when grown on media containing 0.25% or 2.5% lactose are inhibited by a *Streptococcus (strain CA331)* isolated from the caecal contents of adult chickens (Hinton *et al.* 1992). Glass *et al.* (1992) considered the fate of *E. coli 0157:H7* as affected by pH or sodium chloride and in fermented, dry sausage. They concluded that it is important to use beef with low or no *E. coli 0157:H7* in sausage batter because when 10^4/g are present initially, the organism can survive fermentation, drying and storage of fermented sausage, regardless of whether an added starter is present. Clearly this is an area worthy of further investigation, since *E. coli 0157:H7* has been associated with ground beef in a number of food poisoning outbreaks.

IMMOBILIZED CELLS IN MEAT FERMENTATION

Starter cultures must be preserved to retain their cell viability and fermentative activity. Lyophilization is often used for this purpose and Kearney *et al.* (1990) demonstrated that cultures immobilized in alginate beads fermented meat more rapidly than comparable free cell cultures. When starter cultures are inoculated into meat immobilized cells will have the advantage of the additional nutrients in the bead to promote rapid activation of the cells upon rehydration. Schillinger and Lucke (1989) observed a greater level of antagonistic effect on solid media by comparison with liquid culture but it remains to be determined if immobilized LAB also show enhanced antagonistic effects during meat fermentations.

CONCLUSIONS

It is clear that the presence of the normal spoilage or general bacterial biota on meat or poultry cannot guarantee the elimination of pathogens in MAP or fermented meat or poultry products. Initial numbers of pathogens or general biota will have a significant effect on the outcome of the interactions. Modified Atmosphere Packaging may in some instances actually select for certain pathogens. Since some pathogens may grow at low temperature, it is clear that strict quality assurance of meat and poultry products is essential.

REFERENCES

BARAN, L. and STEVENSON, K.E. 1975. Survival of selected pathogens during processing of a fermented turkey sausage. *J. Food Sci. 40*, 618–620.

BLACK, R.E., LEVINE, M.M., CLEMENS, M.L., HUGHES, T.P. and BLASER, M.J. 1988. Experimental *Campylobacter jejuni* infections in humans. J. Infect. Diseases *157*, 472–479.

BLASER, M.J., TAYLOR, D.N. and FELDMAN, R.A. 1984. Epidemiology of *Campylobacter infections*. In *Campylobacter Infections in Man and Animals* (J.P. Butzler, ed.) pp. 143–161, CRC Press, Boca Raton, Fla.

BODDY, L. and TEMPANNY, J.W.T. 1992. Ecological concepts in food microbiolgy. Society for Applied Bacteriology Symposium Series, Number 21. In *Ecosystems: Microbes: Food* (R.G. Board, D. Jones, R.G. Kroll, and G.L. Pettipher, eds.) pp. 23S–39S.

BROWN, M.H. 1982. Introduction. In *Meat Microbiology* (M.H. Brown, ed.) pp.1–11, Appl. Sc. Publishers Ltd., New York.

BRYAN, F.L. 1968. What the sanitarian should know about Staphylococci and Salmonellae in non-dairy products. Staphylococci. J. Milk and Food Technol. *31*, 131–140.

BRYAN, F.L. 1980. Poultry and meat products. In *Microbial Ecology of Foods Vol. 2* Food Commodities (Eds. International Commission on Microbiological Specifications for Foods) pp. 410–469. Academic Press, New York.

BUNCIC, S., PANNOVIC, I. and RADISIC, D. 1991. The fate of *Listeria monocytogenes* in fermented sausage and in vacuum packaged frankfurters. J. Food Protection *54*, (6) 413–417.

COOPER, G.I. 1994. Salmonellosis — infections in man and the chicken: pathogenesis and the development of live vaccines — a review. *Veterinary Bulletin 54*, (2) 123–143.

DAINTY, R.H., SHAW, B.G. and ROBERTS, T.A. 1983. Microbial and chemical changes in chill-stored red meats. In *Food Microbiology: Advances and Prospects* (T.A. Roberts and F.A. Skinner, eds.) pp. 151–178.

DALY, C., LACHANCE, M., SANDINE, W.E. and ELLIKER, P.R. 1973. Control of *Staphylococcus aureus* in sausage by starter cultures and chemical acidulation. J. Food Sci. *38*, 426–430.

DOYLE, M.P. and JONES, D.M. 1992. Food borne transmission and antibiotic resistance of *Campylobacter jejuni*. In Campylobacter jejuni *Current Status and Future Trends*.(I. Nachamkin, M.J. Blaser and L.S. Tompkins, eds.) pp. 45–49, American Society for Microbiology.

DOYLE, M.P. and SCHOENI, J.L. 1984. Survival and growth characteristics of *E.coli* associated with haemorrhagic colitis. Appl. Environ. Microbiol. *28*, 855–856.

DOYLE, M.P. and SCHOENI, J.L. 1989. Isolation of *E.coli 0157:H7* from retail fresh meats and poultry. Appl. Environ. Microbiol. *53*, 2394–2396.

FRANCO, D. 1988. *Campylobacter* species: Considerations for Controlling a Foodborne Pathogen. J. Food Protection *51*, (2) 145–153.

GILL, C.O. 1986. The Control of Microbial Spoilage in Fresh Meats. In *Advances in Meat Research* (Pearson and Dutson, eds.) pp. 49–81. Van Nostrand Reinhold/AVI, New York.

GILL, C.O. and REICHEL, M.P. 1989. Growth of the cold tolerant pathogens *Yersinia enterocolitica, Aeromonas hydrophila* and *Listeria monocytogenes* on high pH beef vacuum packaged under 100% CO_2. Food Microbiol. *6*, 223–230.

GLASS, K.A. and DOYLE, M.P. 1989. Fate of and thermal inactivation of *Listeria monocytogenes* in beaker sausage and pepperoni. J. Food Protection *52*, 226–231, 235.

GLASS, K.A., LOEFFELHOLZ, J.M., FORD, J.P. and DOYLE, M.P. 1992. Fate of *E.coli 0157:H7* as affected by pH or sodium chloride and in fermented, dry sausage. Appl. Environ. Microbiol. *58*, (8) 251–256.

GOEPFERT, J.M. and CHUNG, K.C. 1990. Behaviour of *Salmonella* during the manufacture and storage of fermented sausage. J. Milk Food Technol. *33*, 85–91.

GRAU, F.H and VANDERLINDE, P.B. 1990. Growth of *Listeria moncytogenes* on vacuum-packaged beef. J. Food Protection *53*, (9), 739–741.

GRAU, F.H. and VANDERLINDE, P.B. 1992. Occurrence, numbers and growth of *Listeria monocytogenes* on some vacuum packaged processed meats. J. Food Protection *55*, (1) 4–7.

HANNA, M.O., ZINK, D.L., CARPENTER, Z.L. and VANDERZANT, C. 1976. *Yersinia enterocolitica* like organisms from vacuum packaged beef and lamb. J. Food Sci. *41*, 1254–1256.

HARRIS, N.V., WEISS, N.S. and NOLAN, C.M. 1986. The role of poultry and meats in the etiology of *Campylobacter jejuni/coli* enteritis. Am. J. Public Health *76*, 1560–1566.

HINTLIAN, C.B. and HOTCHKISS, J.H. 1987. Comparative growth of spoilage and pathogenic organisms on Modified Atmosphere Packaged Cooked Beef. J. Food Protection *50*,(3) 218–223.

HINTON, A. CORRIER, D.E. and DeLOACH, J.R. 1992. *In vitro* inhibition of *Salmonella typhimurium* and *E.coli 0157:H7* by an anaerobic gram-positive coccus isolated from caecal contents of adult chickens. J. Food Protection *55*, 162–166.

IBRAHIM, A. and MacRAE, I.C. 1991. Isolation of *Yersinia enterocolitica* and related species from red meat and milk. J. Food Sci. *56*, (6) 1524–1526.

JOHNSON, J., DOYLE, M. and CASSENS, R.G. 1990. *Listeria monocytogenes* and other *Listeria spp.* in meat and meat products. A review. J. Food Protection *53*,(1), 81–91.

JOHNSON, J.L., DOYLE, M.P., CASSENS, R.G. and SCHOENI, J.L. 1988. Fate of *Listeria monocytogenes* in tissues of experimentally infected cattle and in hard salami. Appl. Environ. Microbiol. *54*, 497–501.

KEARNEY, L., UPTON, M. and McLOUGHLIN, A. 1990. Meat fermentations with immobilised lactic acid bacteria. Applied Environ. Microbiol. *33*, 648–651.

KLEINEN, M. and UNTERMANN, F. 1990. Growth of pathogenic *Yersinia enterocolitica* strains in minced meat with and without protective gas with consideration of the competitive background flora. Int. J. Food Microbiol. *10*, 65–72.

LECHOWICH, R.L. 1988. Microbial challenges of refrigerated foods. Food Technol. *42*,(12) 84–89.

LEWUS, C.B., KAISER, A. and MONTVILLE, T.J. 1991. Inhibition of foodborne pathogens by bacteriocins from lactic acid bacteria isolated from meat. Appl. Environ. Micro. *57*,(6) 1683–1688.

LILLARD, H.S. 1971. Occurrence of *Clostridium perfringens* in broiler processing and further processing operations. J. Food Sci. *34*, 1008-1110.

LINDGREN, S.E. and DOBROGOSZ, W.J. 1990. Antagonistic activities of Lactic Acid Bacteria in food and feed fermentations. FEMS Microbiology Reviews *87*, 149-163.

LUCKE, K. and ROBERTS, T.A. 1992. Control in meat and meat products. In *Clostridium botulinum. Ecology and Control in Foods* (A.H.W. Hauschild and K.L. Dodds, eds.) pp. 177-209, Marcel Dekker, New York.

LUITEN, L.S., MARCHELLO, J.A. and DRYDEN, F.D. 1982. Growth of *Salmonella typhimurium* and mesophilic organisms on beef steaks as influenced by type of packaging. J. Food Protection *45*, 263-267.

MARSHALL, D.L., WIESE-LEIGH, P.L., WELLS, J.H. and FARR, A.J. 1991. Comparative growth of *Listeria monocytogenes* and *Pseudomonas fluorescens* on precooked chicken nuggets stored under modified atmospheres. J. Food Protection *54*,(11) 841-843, 851.

McLOUGHLIN, A.J. and CHAMPAGNE, C.P. 1995. Immobilised cells in meat fermentation. (In Press)

METAXOPOULOS, J., GENIGEORIS, M., FANELLI, M.J., FRANTI, C. and COSMA, E. 1981a. Production of Italian dry salami. 1. Initiation of Staphylococcal growth in salami under commercial manufacturing conditions. J. Food Protection *44*, 347-352.

METAXOPOULOS, J., GENIGEORIS, M., FANELLI, M.J., FRANTI, C. and COSMA, E. 1981b. Production of Italian dry salami: Effect of starter culture and chemical acidulation on Staphylococcal growth in Salami under commercial manufacturing conditions. Appl. and Environ. Microbiol. *44*, 863-871.

MOSSEL, D.A.A. 1983. Essentials and perspectives of the microbial ecology of foods. In *Food Microbiology: Advances and Prospects* (T.A. Roberts and F.A. Skinner, eds.) pp. 1-45. Academic Press, London.

MOSSEL, D.A.A. and THOMAS, G. 1988. Cited by B.M. Lund and S.H.W. Notermans. In *Clostridium botulinum* Ecology and Control in Foods (A.H.W. Hauschild and K.L. Dodds, eds.) pp. 279-303, Marcel Dekker, New York.

NORBERG, P. 1981. Enteropathogenic bacteria in frozen chicken. Appl. Environ. Microbiol. *42*, 32-34.

PHEBUS, R.K., DRAUGHON, F.A. and MONET, J.R. 1991. Survival of *Campylobacter jejuni* in modified atmosphere turkey roll. J. Food Protection *54*,(3) 194-199.

POST, L.S., LEE, D., SOLBERG, M., FIRGANG, D., SPECCHIO, J. and GRAHAM, C. 1985. Development of botulinal toxin and sensory deterioration during storage of vacuum and modified atmosphere packaged fish fillets. J. Food Sci. *50*, 990-996.

REYNOLDS, G.N. and DRAUGHON, F.A. 1987. *Campylobacter jejuni* in vacuum packaged processed Turkey roll. J. Food Protection *50*,(4) 300–304.

RIEMANN, H., LEE, W.H., and GENIGEORIS, C. 1972. Control of *Clostridium botulinum* and *Staphylococcus aureus* in semi-preserved meat products. J. Milk and Food Technol. *35*, 514–523.

ROBINSON, D.A. 1981. Infective dose of *Campylobacter jejuni* in milk. Br. Med. J. *282*, 1584.

SALEH, M.A. and ORDAL, I.J. 1955. Studies on growth and toxin production of *Clostridium botulinum* in a precooked frozen food. ii. Inhibition by Lactic Acid Bacteria. Food Res. *20*, 340–350.

SCHILLINGER, U. and LUCKE, F.K. 1989. Antibacterial activity of *Lactobacillus sake* isolated from meat. Appl. Environ. Microbiol. *55*, 1901–1906.

SCHILLINGER, U., KAYA, M. and LUCKE, F.K. 1991. Behaviour of *Listeria monocytogenes* in meat and its control by a bacteriocin producing strain of *Lactobacillus sake*. J. Appl. Bacteriol. *70*, 473–478.

SMITH, J.L. and PALUMBO, S.A. 1981. Microorganisms as food additives. J. Food Protection *44*, (12) 936–955.

STEIR, R.F., BELL, L., ITO, K.A., SHAFER, B.D., BROWN, L.A., SEEGER, M.L., ALLEN, B.H., PORCUNA, M.N. and LERKE, P.A. 1981. Effect of modified atmosphere storage on *Clostridium botulinum* toxigenesis and the spoilage microflora of salmon fillets. J. Food Sci. *46*, 1639–1642.

TRAN, T.T, STEPHENSON, P. and HITCHINS, A. 1990. The effect of aerobic mesophilic microflora levels on the isolation of inoculated *Listeria monocytogenes* strain LM82 from selected foods. J. Food Safety, *10*, 267–275.

WHO 1980. Enteric infections due to *Campylobacter, Yersinia, Salmonella* and *Shigella*. Bull. Wld. Hlth. Org. *58*, 519.

WIMPFHEIMER, L., ALTMAN, N.S. and HOTCHKISS, J.H. 1990. Growth of *Listeria monocytogenes* Scott A, Serotype 4 and competitive spoilage organisms in raw chicken packaged under modified atmospheres and in air. International J. Food Microbiology *11*, 205–214.

CHAPTER 13

DEVELOPMENT AND DISSEMINATION OF CONSUMER INFORMATION TO ENHANCE SAFE HANDLING OF MEAT AND POULTRY PRODUCTS

SHARIN SACHS

Associate Director
Information and Legislative Affairs
USDA Food Safety and Inspection Service
Room 1175, South Building, Washington, DC, 20250

ABSTRACT

The ultimate goal of consumer information on labels and consumer education on food safety is to prevent disease and promote health. This is being achieved through an education program carried out by the Food Safety Inspection Service (FSIS) of the United States Department of Agriculture (USDA). One activity is the 'Hotline direct' consumer service. This, and other educational programs presently in operation are carried out in collaboration with USDA agencies. There is presently a cooperative arrangement with the Health Department on nutritional labelling and food safety education. The educational aspects of food safety and food handler training are being integrated into hazard analysis of critical control points (HACCP)-type operations. The use of HACCP in food safety education is discussed.

INTRODUCTION

HACCP, as an effective means for producing safe food, has been increasingly applied in many areas of the food industry during the last thirty years. Although this approach has been used primarily in food processing, the National Academy of Sciences have suggested that it can be effectively implemented from "farm to fork". Recognising the critical need to educate those who prepare and process consumer food as part of such a "holistic" approach, the Food Safety and Inspection Service (FSIS) has begun to apply the HACCP concept to food safety education. Working with the Extension Service of the US Department of Agriculture (USDA), the Food and Drug Administration (FDA), professional and trade associations, as well as consumer groups, the FSIS has, during recent years, established many information and education projects aimed at preventing disease and promoting consumer health.

Education in Context

Educational activities form only part of the Agency's (FSIS) work. Most of the resources (i.e., about 85%), are directed into its primary regulatory mechanism, nationwide inspection of meat and poultry slaughtering and processing plants.

Other programs of significance in consumer health and safety include the prior label approval scheme, designed to prevent labelling which could confuse or mislead consumers, and the Science and Technology Program which draws on the research capabilities of the USDA's Agricultural Research Service and the broader scientific community.

Within FSIS, the primary responsibility for food safety education and policy communication rests with staff in Information and Legislative Affairs, although educational activities are in practice closely intertwined with the other work of the Agency. Staff from a number of areas in FSIS are involved in the development and delivery of information on the USDA's Toll-Free Meat and Poultry Hotline.

The main business of the Hotline is direct consumer service. In 1993 more than 130,000 members of the public called the Hotline, with the majority of calls being enquiries about basic safe food handling practices. Callers receive a warm, individualised service which has important positive effects on food handler behaviour, and as a side benefit, makes a positive contribution to "government" credibility. The Hotline is largely staffed by a small number of part-time professional dietitians, nutritionists, microbiologists and home economists, who have experience in industry, government and clinical settings. This pattern of part-time employment means that professionals with a very wide range of management and technical expertise can be made available with the flexibility to match the service to the ebb and flow of demand.

Hotline enquiries are analyzed in terms of patterns and trends, and at times, callers are directly surveyed, helping the Agency to more appropriately focus food safety education efforts. A computerized system for tracking and analyzing Hotline calls is currently available. Information gained is distributed to a larger network of information multipliers, including the USDA's Extension Service. Increasingly, this process includes "electronic accessibility" which is beginning to replace paper based distribution systems.

As well as these individual advice activities, the Hotline also carries a menu of recorded messages on many food safety topics, including basic safe food handling techniques. The Hotline can also be involved in Emergency Program activities. For example, on the rare occasions when FSIS Emergency Programs staff determine that a Class I product recall is warranted, to prevent harm to the public, the Hotline can carry a message about the recall, based on the Agency's press release. The recording, available 24 h a day, advises consumers not to eat the product in question, but to return it to the store. The message will clearly identify the product and provide customer education by describing the nature and cause of the problem. This information service enables consumers who have only received part of the information through normal media channels to call and get comprehensive and accurate information on a product they may be holding.

As well as the "direct" consumer education and information services associated with the Hotline, FSIS has a number of more strategic functions. The Food Safety Education Office focuses on long-term educational planning, implementation and evaluation in consumer food safety education. Current priorities include the production and dissemination of information and support material for "at-risk" groups of consumers, such as children, the immuno-compromised and those with chronic health conditions for whom foodborne illness could have more serious consequences.

A considerable amount of this work is carried out in collaboration with other organisations which have similar educational goals. Such cooperation is very important in increasing overall effectiveness. The American public is continually barraged with advice and information. In order to be heard, government agencies must speak with a single clear voice. The involvement of the professional community, consumer groups and industry can maximize the educational impact and reduce waste and duplication of effort. Productive cooperation in consumer education requires the involvement of career specialists and mid-level managers in all phases of planning, implementation and evaluation. This establishes the effective long-term working relationships needed to turn good policy ideas into real accomplishments.

Nutritional Labelling

A good example of such collaboration is the United States Government-wide initiative to require nutritional labelling on all foods. Within this activity, FSIS along with other USDA agencies and the FDA have developed a strong base of support for cooperation in food safety education. In the United States, the FSIS regulates labelling of meat and poultry products while the FDA of the Health and Human Services Department regulates labelling of all other foods. Thus when the FDA was required to mandate nutritional labelling by congressional legislation, FSIS followed suit in the interest of harmonizing food labelling requirements. Therefore, virtually all processed foods, including foods imported from other countries are now required to carry nutritional facts to support informed dietary choices. Because dietary patterns are linked to disease prevention and health promotion, it is hoped that the "nutrition facts" panel, by providing accurate information on the nutrients in a food, will complement current ingredient information to assist consumers to make sensible dietary choices.

In a similar way to the cooperation on regulatory agencies, FDA and FSIS have worked together on a number of other aspects, including educational aspects of consumer nutritional information and food safety education. The nutritional educational strategy and materials were planned together and the resource development and dissemination costs shared. An educational database and a loose "exchange" have been developed to share and distribute information about label education efforts among nutritional community, nonprofit and private sector educators and the public. Thus, the Food Labelling Education Information Center at the National Agricultural Library can assist nonprofit educational organizations with information on previously developed materials, contacts with other organisations with similar objectives, e.g. trade associations, and possible funding sources. This process can help identify gaps in provision and focus nutrition education activities to ensure comprehensive coverage, rather than unnecessary duplication of effort. Similarly, the cooperative experience developed in relation to nutritional labelling education has been applied in food safety education.

Integration of HACCP

In the mid-1980's, staff from Information and Legislative Affairs (ILA) had the opportunity to learn about HACCP, as part of a broader initiative involving their scientific, technical and inspection colleagues. In the early stages the educational program had a primarily public relations approach, using print materials and educational campaigns to convey common sense points about food safety. Such an approach, based on technically accurate information, gave good publicity and did in fact provide educationally effective materials, some of which won national communication awards.

Yet in terms of integrating a comprehensive HACCP perspective to food handler education and behavior modification, and pursuing the strategic objective of preventing foodborne illness, questions remained, e.g.,

(1)   Which food handling practices are most likely to increase the risk of foodborne illness, if they are not carried out properly?

(2)   What do consumers know and practice in terms of safe food handling?

The need for accurate answers to these questions matched the concerns emerging from the National Academy reports to FSIS in 1985 and 1987. In particular, one of these reports noted that "the best inspection system would also encourage individual consumer responsibility for food safety, through effective education programs aimed at the general public, beginning with the school-aged population".

The ideas of HACCP in food handler training and individual consumer responsibility were integrated into the Margin of Safety Project carried out during the late 1980's which aimed overall to collate and review current provision and distribution of food safety education and knowledge, as well as identify themes which required greater attention.

The major activities carried out in the Margin of Safety Project included:

(1)   Tabulation of all available consumer targeted statements from educational materials and services which directly related to food handler behaviour. Seventy six such behaviors were identified and grouped into five "zones" which were under the direct control of the home food handler, i.e., Acquisition, Home Storage, Initial Cooking, Handling and Leftovers.

(2)   Evidence from the scientific literature to directly support the advice statements was sought. Not much was found because the advice statements were based on and extrapolated from "microbiological common sense" about time-temperature abuse, cross-contamination, acidity, water activity, fat content of the food, etc.

(3)   A literature search was conducted, first on HACCP, then on HACCP applied to education. However, not much was available. Epidemiological studies were found based on Centers for Disease Control data on (1) factors contributing to foodborne illness and (2) foods most commonly involved in outbreaks of illness, over time. Most of this work had been completed by Dr. Frank Bryan during his tenure at the Centers for Disease Control and Prevention (CDC). The data found were old and dealt only with a fraction of outbreaks in which epidemiological 'detective work' had successfully confirmed food vehicles, the specific pathogen that caused illness, and the sequence of events that led to illness. Nevertheless, patterns over time were clearly visible.

Examination of epidemiological data sets allowed the identification of a number of areas of concern, some of which had not been previously or adequately recognised. The researchers were able to identify common food handling mistakes by looking for patterns over time. Thus, although undercooking had been previously known as a problem area, the analysis showed that subsequent slow uneven cooling of cooked foods was an additional common factor in outbreaks of illness in the United States. Identification of this risk allowed greater emphasis to be given to the importance of rapid even cooling as a component of effective consumer related HACCP educational activities.

Similarly, epidemiological data identified practices in relation to leftovers as an area which was highly vulnerable to abuse and was therefore a priority for food safety education.

Despite such valuable progress, the working party still had many questions for which no clear answers could be found. For example, which specific food handling mistakes were most likely to lead to foodborne illness? Could the relative risks of various mistakes be compared? Could the cumulative risks be assessed?

To progress the work, Delphi forecasting studies were carried out with a group of nationally respected microbiologists and members of the National Advisory Committee on Microbiological Criteria for Foods. These experts were asked to rate the risk to a high-risk consumer in not performing a specific food handling behavior correctly because, as noted in the project report "under the HACCP approach, the importance of a control point is determined largely by the consequences of not acting at the control point". The "critical control points" are those where failure to take corrective action is most likely to jeopardize the outcome of the process".

Integration of the views of the experts, with the analysis of epidemiological data, gave some answers to the basic question "what do food preparers need to know about safe food handling"? Other questions however remained: what do home food preparers understand about safe food handling? and what safe food handling behaviors do home food preparers actually practice? Further survey activities, including a trawl of consumer surveys, interpreted in collaboration with consumer affairs experts, have been consolidated to form part of the Margin of Safety Project Report. The final report presents a number of useful conclusions, recommendations and themes that deserve more emphasis in food safety education. Today, requests are still being received for these publications, and it is believed they have influenced thinking about food safety education in the United States.

Two basic publications have been developed which remain as mainstays of food safety education. One is the "Quick Consumer Guide to Safe Food Handling" and the other, "Preventing Foodborne Illness" is a reference for Extension agents, local public health officials, and others who work directly with the public and want to understand the hows and whys of foodborne illness. "Preventing Foodborne Illness" might be thought of as a teacher's guide, since it has much more detail and background information than the "Quick Consumer Guide."

Perhaps the most valuable outcome from the Margin of Safety Project was the stimulation of cooperation between those who share common objectives in food safety education. Such collaboration includes:

(1)   Canadian food safety educators adapted the "Quick Consumer Guide" and improved upon it by developing a freestanding display for the brochures.

(2)   A working group of Extension agents reviewed and helped pretest the Quick Consumer Guide and Preventing Foodborne Illness, and with Extension and the FDA, the fourth national satellite teleconference on food safety issues for local public health officials is being planned. The previous interactive "national conference calls" have been very well-received.

(3)   Cornell University faculty in the last three years supervised an updated survey of consumer knowledge and behavior of safe food handling, using the "survey of surveys" as a starting point.

(4)   A major supermarket trade association developed a HACCP-based training guide for delicatessen workers.

(5)   Extension food safety specialists developed HACCP training programs for local public health officials and food service trainers.

(6)   The CDC joined FSIS, FDA and trade associations in a listeriosis education program directed at immuno-compromised persons and others most at risk of serious health consequences from listeriosis.

One of the most significant collaborations has been the recent development of the USDA-FDA Foodborne Illness Education Information Center at the National Agriculture Library which will work with those involved in consumer education and provide a sharper focus on educational research in this area. In particular the Foodborne Illness Education Information Center will address the issue of effectively educating those responsible for preparing foods as a fundamental component of meaningful HACCP systems, i.e., applying adequately rigorous analysis to what is a critical control point in reducing foodborne illness. Therefore, plans include periodic assessment of the extent to which food preparers read, understand and follow safe handling instructions on labels, leading to lasting changes in behaviors which may put them and the final consumer of the foods at risk.

Such activities hold great potential in focusing, implementing, supporting and assessing food safety education efforts. Nevertheless, the evaluation of educational efforts and effects must be as scientific as other aspects of a HACCP approach to food safety and the prevention of foodborne illness. The safe handling label may, for example, prove to be one of the most effective educational mechanisms for improving food safety. However, this must be established rather than assumed, by measuring the effectiveness of safe handling label instructions and other educational efforts in professionally credible ways, relating them to trends in foodborne illness, as is the case with other pathogen reduction efforts.

In summary, if educational analysis, planning and evaluation are conducted appropriately, using the scientific process, a positive side effect may be improved public confidence in food producers and government regulators. Public understanding and trust can only be earned through open, genuine, two-way communication and it must be re-earned every day. Whether educators are in the United States or Ireland, they are far more capable of achieving their educational goals and maintaining trust if they work together.

CHAPTER 14

LINKING MICROBIOLOGICAL CRITERIA FOR FOODS WITH QUANTITATIVE RISK ASSESSMENT

JOAN B. ROSE[1], CHARLES N. HAAS[2], and CHARLES P. GERBA[3]

[1]University of South Florida, Dept. Marine Sciences, St. Petersburg, FL 33701
[2]Environmental Studies Institute, Drexel University, Philadelphia, PA 19104
[3]Department of Soil and Science, University of Arizona, Tucson, AZ 85721

ABSTRACT

The current annual risk of acquiring a foodborne disease in the United States is estimated at 2.7×10^{-2}. The risk of associated death is estimated at 3.7×10^{-5}. These represent a health care burden $> \$3$ billion. Using a risk assessment model one can identify levels of microbial contamination which may be unacceptable in foods and appropriate controls needed to reduce these levels. Salmonella bacteria continue to represent a large percentage of the identifiable infections. A model developed from human dose-response studies predicts the probability of infection for Salmonella at 7.5×10^{-3} with exposure to a single CFU of the organism. Risks of severity (hospitalization), mortality, reactive arthritides, and mortality in the elderly are estimated at 3.1×10^{-4}, 7.5×10^{-6}, 1.7×10^{-5}, and 2.8×10^{-4}. Exposure to microbial contaminants needs to be evaluated on a single meal basis. For chicken, exposure may range from a single drum stick (38g) to a half broiler (176g) but averages around 80g. For beef between 51 and 85g may be consumed during a single meal. Therefore, methods for monitoring must be able to detect at least 1 CFU/80g. Risks for some pathogenic E. coli are estimated at 1,000 to 10,000 less than Salmonella. Therefore, use of coliforms as indicators needs to be assessed and related to occurrence and survival and regrowth potential of the enteric bacteria of greater public health concern. Because, 20% of the U.S. population may be considered to be in a special population category and at an increased risk of severe outcomes, no more than 20% failure of a standard should be acceptable.

INTRODUCTION

An estimated 24 to 81 million cases of foodborne disease occur per year in the U.S. (Archer and Kvenberg 1985). This suggests an annual risk of foodborne disease as high as 1 in 10 to 3 in 10. The microorganisms associated with almost half of the outbreaks go unidentified, however sometimes the agents are identified through epidemiological and outbreak investigations. From 1984 to 1987, 658 foodborne outbreaks had a defined etiology with 89% caused by bacteria (CDC 1990). Control of foodborne disease has focused on proper food processing and handling for providing safe products and preventing recontamination. Yet, the evaluation of various control approaches relies on the ability to monitor for surrogates of contamination, or product abuse. As the outbreak of *E. coli* in hamburgers resulting in thousands of cases and four deaths has demonstrated, better monitoring and control approaches are needed (CDC 1994).

Monitoring data may be useful for evaluating the control of microorganisms at critical control points. Foods may also be examined directly for specific pathogenic microorganisms. Surveys of food products, particularly unprocessed products have detected, but not quantified pathogenic microorganisms. The health impacts associated with low level contamination of foods have also been difficult to quantitate in terms of disease and mortality in the population. These types of quantitative risk assessments, however, can provide information which could lead to better programs for prioritizing and controlling microbial hazards.

Quantitative risk assessment (QRA) is a valuable tool for estimating adverse effects associated with particular hazards and can be an important component of risk management. It is possible to statistically estimate the probability of a harmful event taking place based on exposure and dose-response models. This type of assessment provides managers with valuable information on the identity and characterization of the risks which can be used to develop appropriate control strategies. Risk assessment approaches have been used for estimating the health impacts of chemical contaminants in fish and shellfish (NRC 1991). However, quantitative evaluation of risks associated with exposure to specific microbial contaminants is a relatively new approach and has been used thus far, for microorganisms in drinking water (Haas *et al.* 1993, Rose *et al.* 1991; Rose and Gerba 1991; and Regli *et al.* 1991), and reclaimed wastewater (Rose and Gerba 1990). More recently, quantitative risk assessment has been used for evaluating potential human health impacts associated with exposure to viral contamination of coastal waters and shellfish (Rose and Sobsey 1993).

The four fundamental steps used in a formal risk assessment are: (1) hazard identification, (2) dose-response determination, (3) exposure assessment and (4) risk characterization. This report describes the development and use of quantitative risk assessment to evaluate the risk of infectious disease from exposure to foods contaminated with *Salmonella* compared to *E. coli*.

APPROACH

Identification of the Microbial Hazards

Hazard identification is accomplished by observing and defining the types of adverse health effects in humans associated with exposure to foodborne agents. These include morbidity ratios, illness severity, and mortality ratios. Epidemiological evidence which links the various diseases with the particular exposure route is an important component of this identification. Evaluation of foodborne outbreak statistics is one approach to determine the relative importance of the various hazards. In a great majority of the outbreaks no etiological agent is identified. Therefore, there is inadequate investigation and reporting to accomplish the first step in the risk assessment (hazard identification). Of, outbreaks having a defined etiology, *Salmonella* is the most significant cause of foodborne outbreaks in the U.S. It contributes 52, 61 and 64% of the outbreaks, cases and deaths (CDC 1990). *Clostridium* and *Shigella* also contribute to more cases (5.3 and 19%, respectively) and deaths (20 and 3.4%, respectively) than the other bacteria.

Escherichia coli is generally considered nonpathogenic but includes several types (enterotoxigenic, enteropathogenic and hemorrhagic) associated with foodborne disease. *E. coli* and *Salmonella* non-typhi cause as many as 2.2 million cases of disease and 2,400 deaths each year; an estimated 25 and 96%, respectively are foodborne (Table 1, Bennet *et al.* 1987).

TABLE 1.
CASES, SEVERITY, AND MORTALITY ASSOCIATED WITH *SALMONELLA* AND
E. COLI INFECTIONS IN THE U.S.
(Bennett *et al.* 1987; Gerba *et al.* 1994; Smith *et al.* 1993; Meyers 1989)

	Salmonella	*E. coli*
Annual number of cases	2,000,000	200,000
Annual number of deaths	2,000	400
% Foodborne	96.0	25.0
Mortality Ratio (%)	0.1	0.2[*]
Severity Ratio[**] (%)	4.1	12.7
% Associated with Reactive Arthriditis	2.3	NA
Mortality Ratio in Nursing Homes (%)	3.8	11.8[*]

[*]   *E.coli* 0157:H7
[**]   Hospitalized cases/Total cases during outbreaks

The disease severity can be defined as the number of hospitalized cases/total number of cases recorded during outbreaks (Gerba *et al.* 1994). These ratios were 4.1% for *Salmonella* and 12.7% for *E. coli*. Perhaps only the more virulent isolates are readily identified during outbreaks and therefore are associated with a greater severity rate. Typhoid, for example had a 79% severity rate associated with 4 recorded outbreaks since 1976. The potential for chronic or more serious sequelae is also important. *Campylobacter, Shigella, Yersinia,* and *Salmonella* are associated with reactive arthritides in 2.3% of the cases and Reiter's syndrome in 0.23% of the cases (Smith *et al.* 1993).

The American population is aging and the elderly are at an increased risk of infectious disease. From 1980 to 2020, the number of individuals over 65 is projected to double from 25 to 50 million (Sammons 1986). The fastest growing segment of the population will be the over-85 age group, which is projected to increase from 2.3 to 7.3 million. Infectious diseases are a major problem in the elderly because the immune function declines with age, antibiotic treatment is less effective because of a decrease in physiological function, and malnutrition is more common (Meyers 1989). As a result, outbreaks of gastroenteritis can be devastating in nursing homes and have a significantly higher mortality compared to the general population (Meyers 1989). Most epidemiological studies concerning a specific agent in the elderly are focused around nursing homes since the impact can be more easily observed in a confined group of individuals. Case fatality ratios for specific enteric pathogens are 10 to 100 times greater in this group than the general population. The elderly experience a higher mortality ratio from enteric bacterial gastroenteritis. The overall case fatality ratio for foodborne outbreaks in nursing homes from 1975 to 1987 was 1.0%, compared to 0.1% for outbreaks at other locations (Levine *et al.* 1991).

Morbidity (those with symptoms of disease compared to individuals who are infected and are excreting the bacteria without exhibiting symptoms) is perhaps more varied for the bacteria than any other group of microorganisms. A study of *Salmonella* outbreaks found 6 to 80% of infected individuals became ill (Table 2, Chalker and Blaser 1988). The illness may be mild diarrhea lasting for a few days or severe gastrointestinal illness.

Dose-Response Models

Dose-response models have been developed for bacteria, viruses and protozoa (Haas 1983; Regli *et al.* 1991 and Rose *et al.* 1991). These models were used to evaluate dose-response data sets for *Salmonella* non-typhi and *E. coli* (Blaser and Newman 1982; Graham *et al.* 1983; Ferguson and June 1952; June *et al.* 1953). These data came from human feeding studies, and infection (bacterial colonization and excretion of the microorganism in the feces) was considered a positive response after exposing volunteers to various doses of

TABLE 2.
MORBIDITY* RATIOS FOR *SALMONELLA* (NON-TYPHI)
(Chalker and Blaser 1988)

Study	Population/Situation	Morbidity (0%)
1	Children/Food Handlers	50
2	Restaurant Outbreak	55
3	College Residence Outbreak	69
4	Nursing Home Employees	7
5	Hospital Dietary Personnel	8
6	Hospital Dietary Personnel	6
7	Nosocomial Outbreak	27
8	Summer Camp Outbreak	80
9	Nursing Home Outbreak	23
10	Nosocomial Outbreak	43
11	Foodborne Outbreak	54
12	Foodborne Outbreak	66
AVG.		41

* Morbidity defined as numbers of people ill/ by the total numbers infected, excreting the agent with
and without symptoms.
Mortality Ratio: 0.001 - 0.002
(Bennett *et al.* 1987)

Salmonella, while diahrreal disease was the endpoint for *E. coli*. Several sets of volunteers were exposed to bacteria at various average concentrations to determine the percentage individuals infected.

In evaluating infectious dose data, ID_{50} (that dose which affects 50% infection of the exposed population) is often used. Alternately, a minimal infectious dose, assumes that each microorganism has an inherent minimal dose in the host, below which an infection will not occur. If the exposures are greater than this threshold of concentrations of microorganisms, a response will be observed in the host. However, the log normal model (which assumes a threshold effect) did not provide an adequate fit to the data sets, despite the fact that the lowest dose tested for *E. coli* was 10,000 CFU and 200 CFU for *Salmonella*.

The single-hit exponential model assumes a two-step process: host exposure followed microbial action. The process includes factors such as host resistance, nonspecific pathogen decay and the presence of noninfective bacteria. Laboratory dose-response studies are generally conducted under conditions where the counts of microorganisms in the administered dose approximate the Poisson distribution. Under these conditions, if one microorganism is sufficient to cause an infection, and if host-microorganism interactions are constant, then the probability of an infection (P_i) resulting from ingestion containing an average number of organisms from a single exposure may be defined by the exponential model. In this equation, r is the fraction (proportion) of microorganisms that are ingested which survive to initiate infections ("host-microorganism interaction probability").

When the dose-response curve was plotted for *E. coli* (P_i versus log(N), the exposure), the slope was generally less than predicted by the exponential model. This difference can be due to the heterogeneity of either the infectivity of the individual microorganisms or the sensitivity of the individual hosts or both. Thus, the model can be generalized by assuming that r is not a constant value, but is described by a beta probability distribution. Haas (1983) generalized this model and with the previously mentioned assumptions, the beta-poisson model was developed as an alternative. The parameters α and β characterize the dose-response curve, and as α increases, the model becomes closer to the exponential model.

These two models provided the best fit to the bacterial data sets (Table 3). The beta-poisson ($P_i = 1-(1+N/\beta)^{-\alpha}$) provides a good fit for the *E. coli* data and the exponential ($P=1-exp-rN$) model was utilized for *Salmonella*. The risk of infection (where P_i is the probability of infection, N is the exposure and α, β and r are values defined by the dose-response curves, specific to the individual organism) could then be estimated based on various exposures.

TABLE 3.
BEST FIT DOSE-RESPONSE PARAMETERS FOR VARIOUS
MICROORGANISMS FROM HUMAN FEEDING STUDIES

Organism	Best Model	Model Parameters
E. coli	beta-poisson	$\alpha = 0.1705$
		$\beta = 1.61 \times 10^6$
Salmonella	exponential	$r = 0.00752$

Models:
$P_i = 1 - (1 + N/\beta)^{-\alpha}$ (beta-poisson model)
$P_i = 1 - exp(-rN)$ (exponential)

P_i = Probability of infection
α, β, r = Parameters defining the dose-response curve
N = Exposure (CFU)

This model defines the risk of infection or risk of disease which is characterized as the ability of the bacteria to colonize the intestinal track as a mathematical probability resulting from a given exposure. Two components of the model aid in characterizing the risk: (1) the level of the microbial contaminant in the food and the amount consumed and (2) the interaction of the particular pathogen and host (defined by the dose-response curve). In this exercise it was assumed that the infection is equal to the disease for *Salmonella* and that there is some constant fraction of individuals where the infection will result in severe outcomes and death. The mortality (death) risk was determined by multiplying the predicted probability of infection by the mortality ratio (Table 1), (i.e., $P_i \times$ mortality ratio = risk of mortality). The risk of severity was the P_i multiplied by the severity ratio and the risk of a chronic condition developing for arthridides was the P_i multiplied by the percentage demonstrated in foodborne outbreaks (Table 1). The risk of mortality in the elderly was estimated by multiplying the P_i by the mortality ratio in the nursing home population.

Exposure Assessment

A single exposure to microbial contaminants in food can result in some risk for humans. This concept has been well documented in foodborne outbreaks where consumption of a single serving has caused illness. Therefore, the attack rates will vary, depending on the concentrations of microorganisms in the food, the distribution throughout the food. The amount consumed and the greater the exposure, the greater the risk. Quantitatively, microbial risks can also be estimated for a single exposure or for multiple exposures $[P_{i, 10\text{exposures}} = 1-(P_i)^{10}$ using an average exposure of N for ten portions or exposures]. Exposure to microbial contaminants needs to be evaluated on a single meal basis, yet there is little information on the average levels of various microbes found in meals consumed. Monitoring data often reports only prevalence and thus distribution frequencies without any quantitative data. Therefore QRA can not be accomplished.

In this example, exposures ranging from 1 CFU to 1000 CFU per serving were used to evaluate the risks. According to nutritional statistics books the average meal for chicken may range from a single drum stick (38g) to a half broiler (176g) but averages around 80g. For beef between 51 and 85g may be consumed during a single meal. In applying the risk assessment model, 100 to 200 g will be assumed as the goal for monitoring the finished product. This portion of food consumption has also been noted in foodborne outbreaks, ranging from 28 to 100g (Blaser and Newman 1982).

RESULTS AND DISCUSSION

Risk Characterization

Risks were estimated for exposure to 1, 10, 100, and 1000 colony forming units (CFU) per meal (Table 4). Risks for *Salmonella* were 10,000 times greater than for *E. coli*. With exposure to even 1 CFU of *Salmonella*, the risks of severity, mortality, reactive arthritides and mortality in the elderly were 300, 7.5, 290 and 170 per million, respectively (Table 5). This is assuming that each infection is associated with an illness. This may overestimate the risk if it is assumed that for every infection only a percentage of the population become ill and only a percentage of those ill will be severely effected. The morbidity ratio averaged 40% for *Salmonella* and therefore the previous risks may be decreased by 60%. This may not be true for the reactive arthritides as the prerequisite for this condition is simply infection.

TABLE 4.
PROBABILITY OF INFECTION ASSOCIATED WITH VARIOUS EXPOSURES TO
SALMONELLA AND *E. COLI*

| Exposure/CFU | Risk of Infection | |
	Salmonella	*E. coli*
1	7.5×10^{-3}	1.0×10^{-7}
10	7.2×10^{-2}	1.0×10^{-6}
100	5.3×10^{-1}	1.0×10^{-5}
1000	9.99×10^{-1}	1.0×10^{-4}

TABLE 5.
ADVERSE HEALTH IMPACTS ASSOCIATED WITH RISKS FROM EXPOSURE
TO *SALMONELLA*

Exposure: 1 colony forming unit

Probability of Infection: $(P_i = 1 - \exp(-0.00752 \times 1))$	7.5×10^{-3}
Probability of Severity: $(P_i \times 0.041)$	3.1×10^{-4}
Probability of Mortality: $(P_i \times 0.001)$	7.5×10^{-6}
Probability of Morality: in Nursing Home Populations $(P_i \times 0.038)$	2.9×10^{-4}
Probability of Reactive Arthritides: $(P_i \times 0.023)$	1.7×10^{-4}

There are an estimated 2 million *Salmonella* and 50,000 *E. coli* cases of foodborne disease and infections per year in the U.S. (Bennet *et al.* 1987). Assuming that there was uniform risk in the U.S. population of 250 million, then the 2 million annual cases of salmonellosis would be predicted by the model if each person was exposed to a minimum of 1 CFU or more once during the year. While for the *E. coli,* levels of 5 CFU for 365 days or 270 CFU for 7 days would predict the 50,000 cases annually.

Outbreak data were compared to the output data of the *Salmonella* model for an assessment of the uncertainty and appropriateness for estimating adverse health outcomes (Blaser and Newman 1982). Exposure estimates associated with outbreaks were placed in the model (N) and the P_i was calculated. The attack rates from various outbreaks were then compared to the P_i. The results are shown in Table 6. A single waterborne outbreak was evaluated as well as outbreaks associated with hamburger, chocolate, ice cream, cheese and ham. Exposures ranged from 28 to 100g at levels between 17 and 10^6 CFU. In the cases where attack rates were available the model predictions were very similar or identical.

TABLE 6.
COMPARISON OF OUTBREAK DATA TO MODEL PREDICTIONS FOR ASSESSMENT OF RISKS ASSOCIATED WITH EXPOSURE TO *SALMONELLA*

Food	Dose CFU	Amount consumed	Attack rates	Predicted P^i
Water	17	1L	12%	12%
Pancretin	200	7 doses	100%	77%
Hamburger	60-230	100 g	?	36-82%
Ice cream	102	1 portion	52%	54%
Cheese	100-500	28 g	28-36%	53-98%
Chocolate	100-250	100 g	?	53-85%
Cheese	10^5	100 g	100%	>99.99%
Ham	10^6	50-100g	100%	>99.99%

CONCLUSIONS

Several conclusions may be drawn from this exercise in risk assessment of food safety. The sensitivity of using coliforms as indicators and the amount of food monitored are not adequate to guarantee safety from infectious agents such as *Salmonella*. Risks from even low levels of exposure, particularly in sensitive populations can be considerable (1/10,000). Finally, the limited outbreak data available suggests that the models can be useful in predicting outcomes of disease.

The role of risk assessment is to provide a scientific basis for making decisions about reasonable risk management approaches. "Reasonable" in this case may be defined as cost-effective, feasible and protective of public health. Certainly protection below disease outbreak levels associated with high exposure to contamination is desirable. Acceptable levels of infectious diseases spread through contaminated food have not been debated by the scientific and public health community. Current risk estimates of *Salmonella* infections per year for the U.S. are 7.1×10^{-3} based on 2×10^6 cases reported to the CDC per year and the U.S. population (Bennett *et al.* 1987). This is estimated from all exposures and is probably an underestimation. The economic impact associated with a mild case of salmonellosis and a case severe enough to see a doctor were $221.00 and $680.00 per case, respectively (Roberts 1988). Therefore savings benefits could be achieved by reducing health care costs associated with even low levels of contamination.

In assessing the potential impact of foodborne disease, it is important to recognize that certain individuals may be at greater risk of serious illness than the general population. Individuals who are at increased risk of developing more severe outcomes from foodborne microorganisms are the very young, the elderly, pregnant women, the immunocompromised, those predisposed with other illnesses, and those with a chemical dependency. As part of a risk assessment approach it is important to consider the significance of the potential for severe consequences in sensitive populations. These populations represent almost 20% of the general population and their numbers are expected to increase. Failure of any standard for providing a safe food product will effect these populations most severely.

Risk modeling is a tool used to estimate adverse outcomes at levels which are below the sensitivity of epidemiological methods. It may be a more cost-effective approach when evaluating impacts. Risk assessment has become a valuable tool for evaluating a variety of health hazards associated with food and water. Risk estimation can provide a useful means for decision makers in the development of standards, treatment requirements, for risk management and risk/benefit analysis. However, the risk assessment approach has only been used on a limited scale for judging the risks associated with waterborne pathogenic microorganisms (Haas *et al.* 1993; Rose *et al.* 1991; Rose and Gerba 1991; Regli *et al.* 1991). Such strategy is needed for foods, particularly given the idea of the zero-tolerance in finished products. No models are currently available for *Bacillus, Clostridium, Campylobacter*, or *Listeria*. Quantitative information is needed and needs to be tied to the ability of the bacteria to grow in the foods. Thus, if one organism had a probability of low infectivity but a rapid growth rate or the ability to grow at refrigeration temperatures, the risk for that organism may be greater. Therefore the probability of growth associated with a failed control or product abuse can be used in the overall risk assessment approach and will lead to better controls.

Many key issues need to be addressed in order for microbial risk assessment to be useful. The acceptable contamination levels associated with product safety need to be agreed upon. The role of co-infections and synergistic and antagonistic factors in foods which may influence the infection should be addressed. Better assessment of the hazards, the severity, the mortality, and the role of secondary spread is needed. Exposure estimates based on meals and servings are needed. However, in particular, risk estimates can not be made without better monitoring data of raw and finished products for specific microbial contaminants. New molecular techniques for identifying microorganisms combined with risk models can make this a viable approach.

REFERENCES

ARCHER, D.L. and KVENBERG, J.E. 1985. Incidence and cost of foodborne diarrheal disease in the United States. J. Food Protec. 48, 887–894.

BENNETT, J.V., HOLMBERG, S.D., ROGERS, M.F. and SOLOMON, S.L. 1987. Infectious and parasitic diseases. Amer. J. Prev. Med. 3 102–114.

BLASER, M.J. and NEWMAN, L.S. 1982. A review of human salmonellosis: I. Infective dose. Rev. Infect. Dis. 4, 1096–1106.

Centers for Disease Control. 1990. Waterborne and Foodborne Disease Outbreaks. Morbidity and Mortality Weekly Report. Vol. 39. pp. 1–57.

Centers for Disease Control. 1991. Summary of Notifiable Diseases, United States. Morbidity and Mortality Weekly Report. Vol. 39, pp. 11–26. Atlanta, GA.

CHALKER, R.B. and BLASER, M.J. 1988. A review of human salmonellosis: III. Magnitude of salmonella infection in the United States. Rev. Infec. Dis. 10, 111-124.

FERGUSON, W.W. and JUNE, R.C. 1952. Experiments on feeding adult volunteers with Escherichia coli 111 B4: a coliform organism associated with infant diarrhea. Amer. J. Hyg. 55, p.222–236.

GERBA, C.P., CRAUN, G.F., HAAS, C.N. and ROSE, J.B. 1994. Waterborne disease: sensitive populations at risk. Submitted for publication.

GRAHAM, D.Y., ESTES, M.K. and GENTRY, L.O. 1983. Double-blind comparison of bismuth subsalicylate and placebo in the prevention and treatment of enterotoxigenic Escherichia coli induced diarrhea in volunteers. Gastroenterol. 85, 1017–1022.

HAAS, C.N. 1983. Estimation of risk due to low doses of microorganisms: a comparison of alternative methodologies. Am. J. Epidemiol. 118, 573–582.

HAAS, C.N., ROSE, J.B., GERBA, C.P. and REGLI, S. 1993. Risk assessment of viruses in drinking water. Risk Analysis.

JUNE, R.C., FERGUSON, W.W. and WORFEL, M.T. 1953. Experiments in feedings adult volunteers with *Escherichia coli* 55 B5: a coliform organism associated with infant diarrhea. Amer. J. Hyg. *57*, 222–236.

LEVINE, W.C., SMART, J.F., ARCHER, D.L., BEAN, N.H. and TAUXE, R.V. 1991. Foodborne disease outbreaks in nursing homes, 1975 through 1987. J. Amer. Med. Assoc. *266*, 2105–2109.

National Research Council. 1991. *Sea Food Safety*. The National Academy Press. Washington, D.C.

MEYERS, B.R. 1989. Infectious diseases in the elderly: an overview. Geriatrics *44*, Suppl. A 4-6.

Morbidity and Mortality Weekly Report. 1994. *Escherichia coli* 0157:H7 outbreak linked to home-cooked hamburger-California, July 1993. *43* (12), 213–216.

REGLI, S., ROSE, J.B., HAAS, C.N. and GERBA, C.P. 1991. Modeling the risk from *Giardia* and viruses in drinking water. J. Amer. Water Works Assoc. Nov. 76–84.

ROBERTS, T. 1988. Salmonellosis control: estimated economic costs. Poultry Sci. *67*, 936–943.

ROSE, J.B. and GERBA, C.P. 1990. Assessing potential health risks from viruses and parasites in reclaimed water in Arizona and Florida. Water Sci. Tech. *23*, 2091–2098.

ROSE, J.B. and GERBA, C.P. 1991. Use of risk assessment for development of microbial standards. Water Sci. Tech. *24*(2), 29–34.

ROSE, J.B., HAAS, C.N. and REGLI, S. 1991. Risk assessment and control of waterborne giardiasis. Amer. J. Pub. Health Assoc. *81*(6), 709–713.

ROSE, J.B. and SOBSEY, M.D. 1993. Quantitative risk assessment for viral contamination of shellfish and coastal waters. J. Food. Protect. *56*, 1043–1050.

SAMMONS, J.H. 1986. AMA insights. J. Am. Med. Assoc. *256*, 3085.

SMITH, J.L., PALUMBO, S.A. and WALLS, I. 1993. Relationship between foodborne bacterial pathogens and the reactive arthritides. J. Food Safety. *13*, 209–236.

CHAPTER 15

THE EPIDEMIOLOGY AND COSTS OF DISEASES OF PUBLIC HEALTH SIGNIFICANCE, IN RELATION TO MEAT AND MEAT PRODUCTS

P.N. SOCKETT

Communicable Disease Surveillance Centre
61, Colindale Avenue
London NW9 5EQ
United Kingdom

ABSTRACT

Meat and meat products are important vehicles of foodborne illness outbreaks in European countries. Salmonellas were the most commonly reported aetiology of infection, although the relative importance of other agents varied. The factors contributing to the increase in food poisoning and salmonellosis in England and Wales related both to foods eaten and their preparation. The implication of foods of animal origin as principle vehicles of infection was strengthened by reports associating these foods with outbreaks of human illness, and reports of salmonella infection in animals and poultry. The current increase in salmonella infection associated with poultry products suggests that reducing infection in, or contamination of poultry could significantly decrease human illness.

The problem of human salmonellosis is multi-factorial. Trends are driven by both intrinsic factors relating to the microbiological quality of the food and standards of preparation, and extrinsic factors, such as ambient temperature, which amplify the intrinsic effects. Many of these factors may be amenable to preventive activities, including programs to reduce infection in animals and poultry and programs to educate the consumer in safe food handling.

The costs of human salmonella infection in England and Wales were estimated to be between £231 million and £331 million in 1988 of which £143 million to £205 million may have been associated with meatborne infection. Reductions in human illness, as modelled by irradiation of poultry meat, could give substantial economic gains.

INTRODUCTION

Foodborne diseases have recently become a matter of great Government and public concern. This interest has been generated by three main factors. First is increased reporting of incidents (i.e., sporadic cases or outbreaks) of foodborne illness. For example, many countries contributing to the World Health Organization (WHO) Surveillance Programme for Europe report increasing numbers of incidents for the years 1985 to 1989 (WHO 1992). Second, is the observed trends in disease aetiology which on the one hand indicate dominance by salmonella infection, which accounts for 75% of over 7000 outbreaks reported by sixteen European countries between 1985 and 1989, where the aetiology of infection was known. On the other hand, a number of agents, newly recognized as foodborne, have emerged as important causes of illness. These include listeriosis, yersiniosis and verocytotoxin-producing *Escherichia coli*. Third is the recognition that the social and economic costs of foodborne disease are significant.

Meat is an important vehicle of infection and diseases of public health significance in relation to meat and meat products fall into two groups. The first includes streptococcal skin infections, anthrax, viral and fungal skin lesions, brucellosis, ornithosis and others which affect those handling carcasses and raw meat. These infections in developed countries are, however, largely confined to workers in the meat and livestock industries, and are usually only of public health importance in these settings. The second group relates to those agents which may cause illness in the consumer, and may affect large numbers of persons. This paper is specifically concerned with this second group.

In order to develop effective preventive programs, it is essential to understand the factors which influence trends in foodborne disease. This paper briefly compares meat-associated foodborne illness in Europe, and analyzes the factors affecting food poisoning, particularly salmonellosis, trends in England and Wales for the period 1960 to 1989. Estimates of the socio-economic costs of salmonellosis in England and Wales, and their use in evaluating the costs and benefits of poultry irradiation are presented.

METHODS

Sources of Data and Data Management

Data about trends in food poisoning and salmonella infection in England and Wales were derived from two main sources. The first, Statutory Notifications and Otherwise Ascertained cases of food poisoning, is collated by the Office of Population Censuses and Surveys (OPCS). The second includes reports of laboratory confirmed salmonella infections to the Public Health Laboratory

Service and detailed reports of outbreaks of food poisoning and salmonellosis to the Communicable Disease Surveillance Centre (CDSC) by laboratories and, since 1980, local authorities in England and Wales.

Information on salmonella incidents in animals was extracted from annual summaries collated by the Ministry of Agriculture, Fisheries and Food (MAFF 1977-1989). Under the Zoonoses Orders 1975 and 1989, an incident may relate to the finding of salmonellae in individual animals or groups of animals or their products or surroundings. Data on salmonella incidents in animals was collated by the Central Veterinary Laboratory prior to 1976 and published by Sojka and Field (1970) and Sojka et al. (1977).

Statistics on the consumption of meat and consumer expenditure on meat products were supplied by the Meat and Livestock Commission (MLC), Milton Keynes, Bedfordshire, or extracted from Meat Demand Trends (MLC 1991). Information on meat retailing patterns and ownership of freezers were extracted from specific articles referenced in the text.

Monthly and annual ambient temperatures for England and Wales, for the period 1960 to 1989, were obtained from the Climatological Department of the London Weather Centre, 284, High Holborn, London.

The spreadsheet package Supercalc 5 produced by Computer Associates International, Slough, Berkshire was used for data management (Computer Assoc. Int. Inc. 1989), and Statistical analysis was performed using GLIM (Payne 1983). The approaches taken for analysis of the data are described under main headings corresponding with those in the results and discussion section.

Food

To identify foods which were particularly associated with salmonella infection, outbreaks reported to CDSC which mentioned a suspect food were analyzed for the period 1960 to 1989. The data were analyzed by the organism identified and type of suspect food. A food was described as "associated" if recorded on the outbreak report as the suspect vehicle of infection. However, for many reports, no microbiological or epidemiological confirmation was given.

Analysis was made of outbreaks associated with manufactured foods. These products have the potential to affect large numbers of people. For this analysis, a manufactured food was defined as "any product of the agricultural industry that had been converted into a consumer product by any process of heat treatment, drying or curing, or fermentation" (T. Baird-Parker, personal communication).

Preparation

Laboratory and local authority reports were analyzed to identify the place where outbreaks occurred. It was assumed that for most outbreaks this also reflected the place where the food was prepared and served if foodborne.

Infection in Animals

To explore the relationship between salmonella infections in humans and animals, analysis was made of trends in reported incidents in cattle, pigs, sheep and poultry from 1960 to 1989.

Patterns of Food Consumption

MLC estimates of annual meat supplies were used to indicate patterns of meat consumption in the UK. It was assumed that the patterns would be the same for England and Wales. These data were used to identify trends in consumption of specific types of meat for the period 1966 to 1989. The possibility that other factors relating to storage and retailing also affected consumption was explored by analysis of consumer expenditure on meats and meat products, availability of domestic freezers and refrigerators, seasonality of purchases and patterns of meat retailing.

Weather

To explore the influence of ambient temperature on salmonella infection, annual totals were regressed against annual average summer temperature based on the months June, July, August, September, for the period 1962 to 1989. The significance of any association between temperature and salmonellosis was tested by comparing the estimated temperature effect with its standard error by a series of linear models (McCullagh and Nelder 1989). Calculations were performed in GLIM. Where outlier observations were recorded, their leverage on the model was estimated; allowance was also made for the underlying increase in salmonella reports over the period.

National Costs in England and Wales in 1988

Evaluation of salmonellosis costs was based on a survey of laboratory confirmed salmonella cases in England and Wales, in 1988 and 1989, utilizing a detailed two-part questionnaire developed to explore the use and costs of environmental health and medical services, lost production to industry and costs to the infected individuals and their family (Sockett and Roberts 1991).

A cost-of-illness approach was used and the principle adopted was to identify opportunity costs. Wherever possible valuations were explicit and care was taken to avoid double counting. Market prices were used in the valuation of resources, e.g., EHOs investigation time and medicines. No attempt was made to apportion capital costs of institutions, e.g., EHDs and hospitals, which are involved in multiple and varied activities. Where market prices were not available, implied values were used as a proxy measure of resources. Where alternative valuations were available, both or a range of values was presented, providing a minimum and maximum cost. Details of the study and the methods of analysis is given elsewhere (Sockett and Roberts 1991; Sockett 1993).

In extrapolating the study results to estimate the annual costs of human salmonella infection in England and Wales, it was necessary to make assumptions about levels of under reporting of illness and the severity of illness and its associated costs. Estimates of the level of under reporting (1/38 of actual cases and 1/6 of treated cases) were obtained from Sockett (1993) and were derived from a MAFF survey of experience of gastroenteritis in the community in 1988 (MAFF 1988), and assumptions about the likely source of infection based on distribution of causal organisms in laboratory reports of gastrointestinal illness to CDSC.

Two approaches were taken to estimate annual costs for England and Wales which incorporated alternative methods for relating costs to severity of illness. The first, minimum estimate, categorized severity according to a subjective assessment, by the case, of how ill he or she was. This assumed that demand for treatment was patient led. A severity index was constructed from answers by individuals to questions about the type and severity of symptoms they experienced. The second, maximum estimate, ranked severity according to the level of treatment demanded and applied average tangible costs for each demand group (those requiring hospital treatment, those requiring at least one home visit by a doctor, those who had surgery consultations only and those who did not see a doctor). Details of how costs were aggregated are given in Sockett (1993).

Modelling the Costs and Benefits of Poultry Irradiation in England and Wales

Irradiation costs were based on the use of a gamma-ray (Cobalt 60) source giving an irradiation dose of 3 kiloGrays (kGy). The costs of irradiation included the capital costs, overheads and operating expenses. In addition, allowance was made for additional transport costs to the poultry industry as well as an annual allowance for product promotion. A weighting of 25%, 50% and 75% of the basic costs was added to incorporate any unforeseen plant and operating costs. The benefits of poultry irradiation are presented as the net present values (1988 prices) discounted over 15 years (the life expectancy of an irradiation plant) using the discount rate of 5% recommended by the Treasury for the appraisal of public works (Department of Health and Social Services 1987) and for a series of assumptions about the expected level of reduction in human salmonellosis.

RESULTS AND DISCUSSION

Meat-Associated Illness in European Countries Between 1985 and 1989

Meat and meat products caused a significant proportion of European outbreaks in which a vehicle was recorded (Table 1). Despite national differences in reporting procedures, variation among specific pathogens among countries, and varied culinary customs, meat and meat products were important vehicles of infection in most countries. The importance of poultry meat was variable and was reported as a major ($\geq 20\%$ outbreaks) source of infection in Albania and the United Kingdom. Salmonellae were the most common agents identified in outbreaks although there were significant reports of other agents such as botulism and trichinosis in other countries.

TABLE 1.
PROPORTION OF OUTBREAKS (WHERE A FOOD WAS IMPLICATED) ASSOCIATED
WITH MEAT IN SEVENTEEN EUROPEAN COUNTRIES: 1985-1989

| Country | Percentage of outbreaks | | Country | Percentage of outbreaks | |
	Meat	(Poultry)		Meat	(Poultry)
Albania	72	(20)	Belgium	53	(7)
Bulgaria	23	(3)	Denmark	32	(7)
Finland	47	(2)	France	29	(NS)
Germany (DR)	58	(12)	Germany (FR)	42	(9)
Israel	25	(8)	Netherlands	15	(4)
Poland	26	(2)	Romania	34	(5)
Spain	5	(NS)	Sweden	34	(7)
Yugoslavia	64	(NS)			
England and Wales	62	(26)	Scotland	72	(43)

NS = not stated
Source: World Health Organization (1992)

Factors Affecting Reporting Trends for Salmonellosis in England and Wales

Trends in food poisoning in England and Wales parallel the incidence of salmonella infection, as indicated by laboratory reported salmonellosis and notifications of food poisoning (Fig. 1). While reported salmonella infection increased from the early 1960s and showed a marked increase after 1985, other bacterial food poisoning accounted for few cases and have either remained steady or have declined since the late 1970s (Sockett 1993).

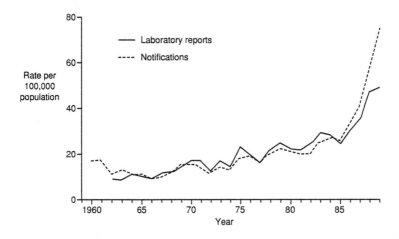

FIG. 1. SALMONELLA REPORTS FROM LABORATORIES AND
FOOD POISONING NOTIFICATIONS. ENGLAND AND WALES:
REPORTING RATES 1960-89

Most laboratory reported salmonella infections were sporadic cases. There is little information about these infections. They may be part of unrecognized outbreaks caused by wide distribution of a common food. Any resulting cases would show a similarly wide distribution and a common vehicle would be difficult to identify (Galbraith 1990). The commonality of serotypes associated with sporadic cases, associated food vehicles and infections in food animals, suggests that many may have been associated with unrecognized outbreaks. The current increase in sporadic human cases corresponds with an increase in *Salmonella enteritidis* infection and increases in outbreaks attributed to poultry and egg containing foods contaminated with *S. enteritidis*, and in *S. enteritidis* infection in poultry. A similar phenomenon was observed with increased *S. typhimurium* infection in humans and cattle in the early 1980s (Sockett *et al.* 1986) and previously in the 1950s and early 1960s (MAFF and PHLS Joint Working Party 1965; Anderson *et al.* 1961; Report of a Working Party of the PHLS 1964).

Food

Evidence linking illness with a particular food was given in only a fifth of salmonella outbreaks compared with higher proportions in outbreaks due to other bacteria (Sockett 1993). Although a wide range of foods were implicated in outbreaks, most were meat and primary agricultural products (Table 2). Most of these products were derived directly from animals, i.e., dairy products and eggs. A few outbreaks were caused by raw vegetables, including spices, bean sprouts, sugar-cane and coconut (Galbraith *et al.* 1960; Anon 1978; Gustavsen and Breen 1984; Anon 1988; O'Mahoney *et al.* 1990).

TABLE 2.
FOOD VEHICLES REPORTED IN OUTBREAKS ENGLAND AND WALES: 1960-1989

Reported Food Item	Salmonella	*Clostridium perfringens*	*Staphylococcus aureus*	*Bacillus* species	Total
		Causative Agent			
Poultry	790	276	74	10	1,150
Beef	66	507	52	7	632
Pork/ham	153	184	94	4	435
Lamb/mutton	8	100	4	1	113
Other or mixed meats, pies, sausages	233	261	117	20	631
Sub-total	1,250	1,328	341	42	2,961
Other/mixed foods	496	121	140	231	988
Total (food reported)	1,746	1,449	481	273	3,949

Poultry was the most commonly implicated meat and the majority of these outbreaks were due to salmonellae. Pork and ham were also commonly associated with salmonella infection as were processed meats, meat pies and pasties. In contrast, salmonella outbreaks associated with beef or sheep meat were, comparatively, less common. The continued association of salmonellosis with meats and meat products indicates the difficulties of controlling contamination of products from animal sources. A series of studies and reports in the UK in the 1950s and 1960s confirmed the chain of infection from animals to humans (Anderson *et al.* 1961; Galbraith 1961; Hobbs 1961; Lee 1974). Specific problem areas identified were feed contamination (Report of a Working Party of the PHLS 1961; Dawkins and Robertson 1967; Harvey and Price 1967; Patterson 1972), control of cross-infection and cross-contamination at the farm, during transport of animals and at the abattoir (Ritchi and Clayton 1951; Harvey and Powell-Phillips 1961; Report of a Working Party of the PHLS 1964; Dixon and Peacock 1965; Hugh-Jones 1969; Payne 1969; Bicknell 1972; Lee *et al.* 1972; Ghosh 1972; PHLS Working Group 1972; Clegg *et al.* 1986; Morgan *et al.* 1987; Wray *et al.* 1987; Wray *et al.* 1990), and prevention of contamination of finished raw products (Galbraith *et al.* 1964; Roberts *et al.* 1975; Lillard 1989).

Relatively few outbreaks were associated with manufactured foods. Salmonellas accounted for 45% of 294 outbreaks due to manufactured foods reviewed for the period 1980 to 1989 (Table 3) (Sockett 1991a). Cooked and processed meats were the most significant vehicles and were particularly associated with salmonella infection. Manufactured products are distributed widely and produced in large quantity and have the potential to affect many people geographically widespread (Gill *et al.* 1983; Rowe *et al.* 1987; Cowden *et al.* 1989). The food may be recalled and the manufacturer may bear considerable economic loss (Sockett 1991b).

TABLE 3.
OUTBREAKS ASSOCIATED WITH MANUFACTURED FOODS*
ENGLAND AND WALES: 1980-1989

Type of Food	Cause of outbreak					
	Salmonella		Other or unknown		Total	
Cooked foods; processed meats; meat pies; potted meats; pâté	107	(4)	42	(2)	149	(6)
Canned foods:						
meats	5		10	(4)	15	(4)
fish and shellfish	1		51	(47)	52	(47)
vegetables	-		1	(1)	1	(1)
Other foods†	19	(3)	58	(4)	77	(7)
Total	132	(7)	162	(58)	294	(65)

*   imported products shown in brackets
†   these include dairy and bakery products and powdered foods

Preparation

The proportion of outbreaks due to salmonella varied with the food preparation site. Single households accounted for the majority (76%) of salmonella outbreaks recorded between 1960 and 1989 (Sockett 1993). Poor handling practices in the domestic kitchen, particularly in the preparation of poultry which has a high risk of contamination, may be an important contributory factor. According to consumer surveys commissioned by MAFF (1988) and J Sainsbury PLC (Anon 1991) many people do not know, or do not employ, basic rules of food hygiene.

General outbreaks, affecting members of more than one household, are shown in Table 4, and reflect the type and scale of catering involved. Food preparation practices associated with general outbreaks fell into five categories. (1) The restaurant type prepares most dishes "to order." (2) The reception type may have a limited menu, but prepares items in advance and stores them ready to serve quickly. (3) The canteen type where salmonella infection was secondary to *Clostridium perfringens* intoxication, indicating the reheating of bulk prepared food. (4) Salmonella infection was the most common agent in outbreaks associated with shops; these commonly implicated sliced cooked meats or pies (Sockett 1991a), and it is likely that cross-contamination via slicers, surfaces and hands were contributory factors. (5) Almost all outbreaks of the farm/dairy type were due to salmonella infection, usually linked with either unpasteurized or inadequately pasteurized milk (Barrett 1989; Sockett 1991c).

TABLE 4.
GENERAL OUTBREAKS DUE TO BACTERIAL FOOD POISONING AND
SALMONELLOSIS.* PLACE WHERE FOOD WAS CONSUMED/BOUGHT
ASSOCIATED WITH 4,827 OUTBREAKS IN ENGLAND AND WALES
FROM 1960 TO 1989

Location where Food was Acquired	Cause of Outbreak			
	Salmonella		Other bacterial	Total
Restaurant, hotel, reception	1,032	(61%)	660†[A]	1,692
Canteen, school	210	(27%)	573†[B]	783
Hospital, institution	1,169	(64%)	665†[C]	1,834
Farm, dairy	222	(99%)	3	225
Shop	219	(75%)	74‡	293

* proportion of outbreaks due to salmonella shown in brackets
† *Clostridium perfringens*: A = 22%; B = 64%; C = 32%
‡ *Staphylococcus aureus*: 20%

Infection in Animals

Trends in salmonella infections in animals since 1960 are shown in Fig. 2. Early peaks in infection in cattle (1970s) and poultry (1960s) were associated with host adapted serotypes, however, more recent trends (1976 onwards) correspond with the recording of poultry, egg, beef and milk products as the foods most often associated with outbreaks in humans. For example, a peak in salmonella infection in poultry in 1980 (Fig. 3), was due to a number of serotypes including *S. agona*, *S. hadar*, *S. senftenberg*, *S. montivideo* and *S. virchow*, whereas a second peak, from 1988, was due largely to *S. enteritidis*. These coincided with peaks in outbreaks of human infection associated with poultry meat in 1980 and poultry meat and egg-containing foods from 1988. Introduction of Statutory Testing of poultry including egg-laying flocks in 1989 under The Poultry Breeding Flocks and Hatcheries (Registration and Testing) Order, 1989 affected trends thereafter.

Trends in incidents in animals were influenced by a small number of serotypes, most also common in humans (Table 5). Thus, 92% of incidents in cattle were associated with the five most common serotypes reported. Proportions in other animal species and poultry varied between 64 and 78%. For each animal species, between one third and over a half of incidents were caused by serotypes also commonly reported in human infections (Table 5).

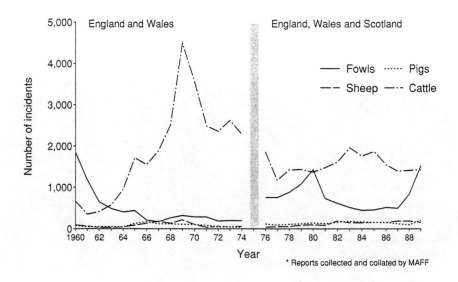

FIG. 2. SALMONELLA INCIDENTS IN ANIMALS* 1960-1989

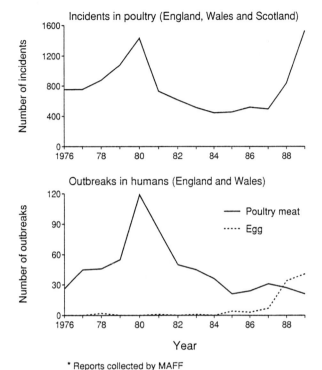

FIG. 3. SALMONELLA: INCIDENTS IN POULTRY* AND OUTBREAKS IN
HUMANS DUE TO POULTRY MEAT AND EGGS+ 1976-1989

TABLE 5.
INCIDENTS OF SALMONELLA INFECTION IN ANIMALS DUE TO THE MOST
COMMON SEROTYPES IN ANIMALS AND HUMANS

	Cattle	Sheep	Pigs	Fowl	Turkeys
Proportion of incidents due to the most common serotypes* in animals	92%	78%	67%	64%	64%
Proportion of incidents due to serotypes which were also common in man†	57%	45%	33%	40%	44%

*   The five most common salmonella serotypes recorded annually in England, Wales and Scotland
   from 1976-1989 under the Zoonoses Order (1975).
†   The five most common serotypes recorded annually in England and Wales.

Patterns of Food Consumption

The type of meat eaten and the way it was presented to the consumer changed over the period reviewed. A continuous increase in poultry production offset a decline in other meats to maintain meat consumption relatively constant. The consumption of chicken is far less seasonal than other meats and is high throughout the summer period when other meat consumption is lowest.

There is also a movement away from fresh meat for immediate consumption (except poultry) towards frozen or processed meat products. Consumption of frozen foods and ready meals grew continuously in the 1980s, corresponding with increased ownership of freezers and a trend towards shopping at supermarkets. Thus in 1980, 23% of poultry meat was purchased in butchers compared with 35% in supermarkets; by 1990 purchases in butchers had declined to 17% whereas supermarkets claimed 52% of purchases (MLC 1991). There is also an increasing tendency for customers to purchase "fresh" meat for freezing rather than meat already frozen (MLC 1991). This trend may increase handling of potentially contaminated products in the kitchen which could result in cross-contamination of other foods.

The importance of poultry as a major vehicle of human salmonellosis therefore relates to four factors. Increased consumption over the period reviewed, the high proportion of contaminated oven-ready birds (D. Roberts 1991), increased use of fresh chicken which may act as a source of contamination in the kitchen, and high consumption, compared with other meats, in the summer months when high ambient temperature increases risks associated with inappropriate transportation, handling, cooking and storage.

Weather

A close relationship between ambient temperature and increased infections is demonstrated by the within-year trends in human salmonella infection. Each year the number of cases increases with the onset of warmer temperatures in May and June and declines in the autumn. This may coincide with seasonal changes in eating habits to those that may be more risky. Long-term trends in salmonella infection indicated that annual trends were also influenced by summer temperatures and that a hot summer resulted in a peak in reporting of salmonellosis (Fig. 4). Multiplication of salmonellas in warm conditions may increase the pool of organisms in the environment and thereby increase the chance that foods will become contaminated. Increased growth rates on contaminated food may increase the likelihood of cross-contamination to other foods. Although the underlying effect of temperature on reporting trends was masked by the overwhelming increase in *S. enteritis* after 1985, the margins for handling error clearly decrease as temperatures increase.

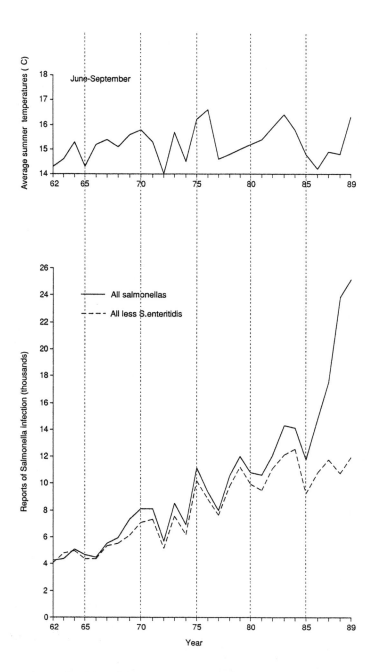

FIG. 4. AVERAGE SUMMER TEMPERATURES AND ANNUAL TOTALS
OF SALMONELLA REPORTS: ENGLAND AND WALES 1962-1989

Factors Affecting Transmission of Salmonella Infection Via Food

These data indicate that factors both intrinsic and extrinsic to food production affect transmission of salmonella infection. Intrinsic factors relate to the type of food and it's preparation. Some foods carry a greater risk of contamination and some preparation practices are more likely to result in cross contamination or microbial growth in food. Extrinsic factors relate, firstly, to the contamination of raw materials, including animals and animal feeds. Such contamination may be exacerbated by the way the material or animal is kept and transported. Secondly, environmental factors such as high ambient temperature can encourage proliferation of salmonellae. Thirdly, changes in consumer demand may affect shopping practices and the supply of particular products towards those having a higher risk of contamination. For example, increasing purchase of fresh meats for freezing will result in increased handling in the kitchen and a greater likelihood of cross-contamination (MLC 1991).

Some of these factors may be amenable to intervention or risk reduction. Thus identifying foods posing a particular risk of contamination, ensuring adequate processing of the food and encouraging high standards of hygiene during handling are likely to reduce the effect of intrinsic factors. Likewise, reducing contamination of raw materials, lowering storage temperatures and encouraging consumers to demand safer products, will reduce the effects of extrinsic factors.

Primary agricultural products can be infected or contaminated directly, whereas after slaughter and during processing and retailing the problems relate either to cross-contamination or proliferation of indigenous organisms. Corresponding preventive activities can prevent or control infection or contamination of the primary product and eliminate contamination of the finished product by heating, chemical treatment or irradiation. Good hygiene practices become increasingly important post processing and are the last line of defense. Obviously measures which reduce contamination at an earlier stage make a failure during final preparation less likely to result in illness.

Economic Impact of Foodborne Illness at the National Level

The very few studies of the economic impact of foodborne illness in the literature indicate the costs of salmonellosis are significant. For example, studies in the then Federal Republic of Germany and Canada in the late 1970s estimated national costs of human salmonellosis of DM 108 million and (Canadian) $84 million (Krug and Rehm 1983; Curtin 1984). Tanya Roberts (1991) has estimated annual costs of $4.8 billion for foodborne disease in the United States, most of which was associated with salmonella, campylobacter and staphylococcal infection.

National Costs in England and Wales in 1988

A detailed study of the costs of human salmonella infection in 1482 individuals in England and Wales for 1988 was published in 1991 (Sockett and Roberts 1991). Extrapolation of the results gave estimated minimum and maximum national costs of illness of £231 million to £331 million for 1988 (Table 6) (Sockett 1993). The range reflected two alternative approaches to estimating costs associated with unreported and mild infections. A further £19 million to £132 million intangible costs were estimated for deaths and for pain and suffering. The cost profile was dominated by costs of production loss to the economy resulting from sickness-related absence from work. Significant costs were also associated with the investigation and medical treatment (public sector costs), and the costs to affected individuals and their families.

A recent study by the Department of Agricultural Economics and Management at Reading University sought to identify the contribution of different food types to trends in salmonellosis in England and Wales (S. Henson, personal communication). Preliminary results have been used here to estimate the proportion of salmonella costs which may be attributed to meat and meat products in England and Wales (Table 6). Assuming that most salmonella infection in England and Wales is foodborne, the costs of meat associated infection were between £143 million and £205 million in 1988. Two thirds of these costs were attributed to poultry meat-borne infection and the remainder to other meat and meat products. These estimates are based on expert opinion and indicate that, given the cost of salmonellosis, the costs and benefits of potential preventive measures should be explored. An example, illustrating the cost-benefit of poultry irradiation is presented below.

Modelling the Cost-Benefit of Poultry Irradiation in England and Wales

The model used to compare the costs of irradiating poultry with the benefits gained from reducing salmonellosis was based on the assumption that all poultry meat was irradiated at ten regional irradiation plants, each with a throughput of 100,000 tonnes per year. This equals the annual poultry consumption in England and Wales (MLC 1991). The capital and maintenance costs of an irradiation plant were adapted from costs given by Yule et al. (1986).

Mulder et al. (1977) showed doses of 2.5 kGy kill salmonellas. At 3 kGy 99.99% of salmonella cells are inactivated under optimal conditions (Kinzel 1991). This is less than half the dose permitted for poultry in the UK under The Food (Control of Irradiation) Regulations, 1990. Even at lower doses, e.g. 0.9 kGy, cell viability in deboned chicken meat at OC was reduced 6.4 log; surviving cells were more heat sensitive and remained so for at least six weeks (Thayer et al. 1991).

TABLE 6.
ESTIMATED COSTS OF MEAT-BORNE SALMONELLOSIS
ENGLAND AND WALES: 1988 PRICES

	Estimated % of salmonellosis caused by meat*	Minimum cost (£ millions)			Maximum cost (£ millions)		
		Total	E†	PS‡	Total	E†	PS‡
Total costs of salmonellosis		231	162	46	331	220	46
Poultry meat	43	99	70	20	142	95	20
Other meat/meat products	19	44	31	9	63	42	9
Total	62	143	101	29	205	137	29

* based on preliminary results of a "Delphi" survey of expert opinion (S Henson, personal communication)
† costs to the Economy from sickness-related absence from work
‡ costs to the Public Sector from the investigation and treatment of infected persons

The benefits of irradiation (costs of salmonellosis less costs of irradiation) are presented in Table 7 (Sockett 1993). Only under the most severe assumptions about irradiation costs and effectiveness was the net benefit negative. That is, benefit was only negative for the highest cost of irradiation (cost + 75%) at the lowest assumption for cases avoided, at minimum cost of salmonellosis. Under these assumptions the point at which benefits broke even with costs occurred when a reduction of between 10 and 25% of cases was achieved for the severest assumptions of irradiation cost, and otherwise at a level of less than a 10% reduction in cases.

The model presented indicates that, even for a modest (25%) reduction in human salmonellosis, the minimum benefits accrued under all assumptions of irradiation cost and discount rates would be substantial (£261 million - £540 million over 15 years) and there would be considerable scope for increasing some elements, e.g., doubling promotional costs, without significantly affecting benefits.

TABLE 7.
FLOW OF BENEFITS (£ MILLIONS)* OF POULTRY IRRADIATION TO REDUCE
HUMAN SALMONELLA INFECTION IN ENGLAND AND WALES: 1988 PRICES

| Irradiation costs† | Flow of benefits at minimum cost of salmonellosis | | | Flow of benefits at maximum cost of salmonellosis | | |
| | Reduction in salmonellosis | | | Reduction in salmonellosis | | |
	10%	25%	50%	10%	25%	50%
Basic cost	94.5	472.1	1,101.5	203.4	744.6	1,6464
Cost + 25%	60.6	438.3	1,067.7	169.6	710.7	1,6126
Cost + 50%	26.8	404.4	1,033.8	135.8	676.9	1,5787
Cost + 75%	-7.0	370.6	1,000.0	102.0	643.1	1,5449

* net present value discounted (5%) over 15 years life of plant

† capital costs of plant (£57,064,560) and land (£650,000) and recurrent expenditure on maintenance, transport and product promotion (£9,136,940) at cost and cost plus 25%, 50% and 75%

REFERENCES

ANDERSON, E.S., GALBRAITH, N.S. and TAYLOR, C.E.D. 1961. An outbreak of human infection due to *Salmonella typhimurium* phage type 20a associated with infection in calves. Lancet *1*, 854–858.

Anon. 1978. Salmonella in vegetables. Brit. Med. J. *1*, 1008.

Anon. 1988. Communicable Disease Report January to March 1988. Community Med. *10*, 250–254.

Anon. 1991. Food safety in the home: survey highlights. J Sainsbury PLC, London.

BARRETT, N.J. 1989. Milkborne disease in England and Wales in the 1980s. J. Soc. Dairy Technol. *42*, 4–6.

BICKNELL, S.R. 1972. *Salmonella aberdeen* infection in cattle associated with human sewage. J. Hyg., Cambridge; *70*, 121–126.

CLEGG, F.G., WRAY, C., DUNCAN, A.L. and APPLEYARD, W.T. 1986. Salmonellosis in two dairy herds associated with a sewage farm and water reclamation plant. J. Hyg., Cambridge; *97*, 237–246.

Computer Associates International Incorporated. 1989. Supercalc version 5. Computer Associates International Incorporated, Slough.

COWDEN, J.M., O'MAHONEY, M., BARTLETT, C.L.R., RANA, B., SMYTH, B., LYNCH, D. *et al.* 1989. A national outbreak of *Salmonella typhimurium* DT 124 caused by contaminated salami sticks. Epidemiol. Infect. *103*: 219–225.

CURTIN, L. 1984. Economic study of salmonella poisoning and control measures in Canada. Working paper 11/84. Food Market Analysis Division. Marketing and economics branch, Agric. Canada, Ottawa.

DAWKINS, H.C.and ROBERTSON, L. 1967. Salmonellas in animal feeding stuffs. Monthly Bull. Min. Health and Pub. Health Lab. Serv. 26, 215–221.

Department of Health and Social Security. 1987. Option appraisal: a guide for the National Health Service. HMSO, London.

DIXON, J.M.S. and PEACOCK, N. 1965. A survey of the contamination with salmonellae of imported Dutch meat in 1960 and 1964. Monthly Bull. Min. Health and Pub. Health Lab. Serv. 24, 361–364.

GALBRAITH, N.S. 1961. Studies of human salmonellosis in relation to infection in animals. Veterinary Rec. 73, 1296–1303.

GALBRAITH, N.S. 1990. The epidemiology of foodborne disease in England and Wales in the 1980s. Outlook on Agric., 19, 95–101.

GALBRAITH, N.S., HOBBS, B.C., SMITH, M.E. and TOMLINSON, A.J.H. 1960. Salmonellae in desiccated coconut. An interim report. Monthly Bull. Min. Health 19, 99–106.

GALBRAITH, N.S., TAYLOR, C.E.D., PATTON, J.L. and HAGAN, J.G. 1964. Salmonella infection in poultry. Med. Officer 111, 354–356.

GHOSH, A.C. 1972. An epidemiological study of the incidence of salmonellas in pigs. J. Hyg., Cambridge; 70, 151–160.

GILL, O.N., SOCKETT, P.N., BARTLETT, C.L.R., VAILE, M.S.B., ROWE, B., GILBERT, R.J., DULAKE, C., MURRELL, H.C. and SALMASO, S. 1983. Outbreak of Salmonella napoli infection caused by contaminated chocolate bars. Lancet 2, 574–577.

GUSTAVSEN, S. and BREEN, O. 1984. Investigation of an outbreak of Salmonella oranienburg infections in Norway, caused by contaminated black pepper. Amer. J. Epidem. 115, 806–812.

HARVEY, R.W.S. and POWELL-PHILIPS, P.W. 1961. An environmental survey of bakehouses and abattoirs for salmonellae. J. Hyg. Cambridge 59, 93–103.

HARVEY, R.W.S. and PRICE, T.H. 1967. The isolation of salmonellas from animal feedingstuffs. J. Hyg. Cambridge 65, 237–244.

HOBBS, B.C. 1961. Public health significance of salmonella carriers in livestock and birds. J. Appl. Bacteriol. 24, 340–352.

HUGH-JONES, M.E. 1969. Epidemiological studies on Salmonella senftenberg II. Infections in farm animals. J. Hyg. Cambridge 67, 89–94.

KINZEL, B. 1991. Breaking the salmonella/chicken connection; irradiation is approved for poultry processing. Dairy, Food and Environ. Sanitation 11, 432.

KRUG, W. and REHM, N. 1983. Nutzen-Kosten-analyse de salmonellose-bekampfung. Schriftenreihe des Bundesministers fur Jugens, Familie und Gesundheit, Berlin.

LEE, J.A. 1974. Recent trends in human salmonellosis in England and Wales: the epidemiology of prevalent serotypes other than *Salmonella typhimurium*. J. Hyg. Cambridge *72*, 185–195.

LEE, J.A., GHOSH, A.C., MANN, P.G. and TEE, G.H. 1972. Salmonellas on pig farms and in abattoirs. J. Hyg. Cambridge *70*, 141–150.

LILLARD, H.S. 1989. Factors affecting the persistence of salmonella during the processing of poultry. J. Food Protect. *52*, 829–832.

MCCULLAGH, P. and NELDER, J.A. 1989. Generalised Linear Models. (2nd ed.). Chapman and Hall, London.

Meat and Livestock Commission. 1991. Recent developments in retail market shares. *Meat Demand Trends*; Feb. 13–17.

Ministry of Agriculture, Fisheries and Food. 1988. Food Hygiene: report on a consumer survey. HMSO, London.

Ministry of Agriculture, Fisheries and Food and the Public Health Laboratory Service, Joint Working Party. 1965. Salmonella in cattle and their feeding-stuffs, and the relation to human infection. J. Hyg. Cambridge *63*, 223–241.

Ministry of Agriculture, Fisheries and Food; Department of Agriculture and Fisheries for Scotland. 1978. Animal Salmonellosis, Annual Summary 1977. Epidemiology Department, Central Veterinary Laboratory, Weybridge.

Ibid. 1979. Annual Summary 1978.

Ibid. 1980. Annual Summary 1979.

Ministry of Agriculture, Fisheries and Food; Department of Agriculture and Fisheries for Scotland; Welsh Office Agriculture Department. 1982. Animal Salmonellosis, Annual Summaries 1980/81. Epidemiology Department, Central Veterinary Laboratory, Weybridge.

Ibid. 1983. Annual Summary 1982.

Ibid. 1984. Annual Summary 1983.

Ibid. 1985. Annual Summary 1984.

Ibid. 1986. Annual Summary 1985.

Ibid. 1987. Annual Summary 1986.

Ibid. 1988. Annual Summary 1987.

Ibid. 1989. Annual Summary 1988.

Ibid. 1990. Annual Summary 1989.

MORGAN, I.R., KRAUTIL, F.L. and CRAVEN, J.A. 1987. Effect of time in lairage on caecal and carcass salmonella contamination of slaughter pigs. Epidemiol. and Infect. *98*, 323–330.

MULDER, R.W.A.W., NOTERMANS, S. and KAMPELMACHER, E.H. 1977. Inactivation of salmonellae on chilled and deep frozen broiler carcasses by irradiation. J. Appl. Bacteriol. *42*, 179–185.

O'MAHONEY, M., COWDEN, J., SMYTH, B., LYNCH, D., HALL, M., ROWE, B., TEARE, E.L., TETTMAR, R.E., RAMPLING, A.M., COLES, M., GILBERT, R.J., KINGCOTT, E. and BARTLETT, L.L.R. 1990. An outbreak of *Salmonella saint-paul* infection associated with beansprouts. Epidemiol. and Infect. *104*, 229–235.

PATTERSON, J.T. 1972. Salmonellae in animal feedingstuffs. Rec. Agric. Res. *20*, 27–33.

PAYNE, C. (ed). 1983. The GLIM system manual for release 3.77. Numerical Algorithms Group, Oxford.

PAYNE, D.J.H. 1969. Salmonellosis and intensive farming. Public Health, London; *84*, 5–16.

Public Health Laboratory Service Working Group, Skovgaard N, Neilson BB. 1972. Salmonellas in pigs and animal feedingstuffs in England and Wales and in Denmark. J. Hyg. Cambridge *70*, 127–140.

Report of a Working Party of the Public Health Laboratory Service. 1961. Salmonella organisms in animal feeding stuffs. Monthly Bull. Mini. Health Pub. Health Lab. Ser. *20*, 73–85.

Report of a Working Party of the Public Health Laboratory Service. 1964. Salmonellae in abattoirs, butchers' shops and home-produced meat, and their relation to human infection. J. Hyg. Cambridge *62*, 283-302.

RITCHIE, J.M. and CLAYTON, N.M. 1951. An investigation into the incidence of *Salmonella dublin* in healthy cattle. Monthly Bull. Min. Health Pub. Health Lab. Serv. *10*, 272-277.

ROBERTS, D., BOAG, K., HALL, M.L. and SHIPP, C.R. 1975. The isolation of salmonellas from British pork sausages and sausage meat. J. Hyg. Cambridge *75*, 173–184.

ROBERTS, D. 1991. Salmonella in chilled and frozen chicken. Lancet *337*, 984–985.

ROBERTS, T. 1991. A survey of estimated risks of human illness and costs of microbial foodborne disease. J. Agribus. *9*, 5–23.

ROWE, B. *et al.* 1987. *Salmonella ealing* infections associated with consumption of infant dried milk. Lancet *2*, 900–903.

SOCKETT, P.N. 1991a. Food poisoning outbreaks associated with manufactured foods in England and Wales: 1980-1989. Comm. Dis. Rep. *1*, R105–R109.

SOCKETT, P.N. 1991b. The economic implications of human salmonella infection. J. Appl. Bacteriol. *71*, 289-295.

SOCKETT, P.N. 1991c. Communicable disease associated with milk and dairy products: England and Wales 1987-1989. Comm. Dis. Rep. *1*, R9–R12.

SOCKETT, P.N. 1993. The economic and social impact of human salmonellosis in England and Wales. A study of the costs and epidemiology of illness and the benefits of prevention. PhD Thesis, University of London.

SOCKETT, P.N. and ROBERTS, J.A. 1991. The social and economic impact of salmonellosis: a report of a national survey in England and Wales of laboratory confirmed infections. Epidemiol. and Infect. *107*, 335–347.

SOCKETT, P.N., ROWE, B. and PALMER, S.R. 1986. Surveillance of *Salmonella typhimurium* phage types: England and Wales 1982-1984. In *Proceedings Second World Congress. Foodborne infections and intoxications.* WHO, Berlin *1*, 138-141.

SOJKA, W.J. and FIELD, H.I. 1970. Salmonellosis in England and Wales 1958-1967. The Veterinary Bulletin; *40*, 515–531.

SOJKA, W.J., WRAY, C., SHREEVE, J. and BENSON, A.J. 1977. Incidence of salmonella infection in animals in England and Wales, 1968-1974. J. Hyg. Cambridge *78*, 43–56.

THAYER, D.W., SONGPRASERTCHAI, S. and BOYD, G. 1991. Effects of heat and ionizing radiation on *Salmonella typhimurium* in mechanically deboned chicken meat. J. Food Protect. *54*, 718–724.

World Health Organization. 1992. World Health Organization Surveillance Programme for Control of Foodborne Infections and Intoxications in Europe. FAO/WHO Collaborating Centre for Research and Training in Food Hygiene and Zoonoses. Berlin.

WRAY, C., TODD, N. and HINTON, M.H. 1987. The epidemiology of *Salmonella typhimurium* infection in calves: excretion of *S. typhimurium* in the faeces of calves in different management systems. Veterinary Rec. *121*, 293–296.

WRAY, C., TODD, N., MCLAREN, I., BEEDELL, Y. and ROWE, B. 1990. The epidemiology of salmonella infection in calves: the role of dealers. Epidemiol. Infect. *105*, 295–305.

YULE, B.F., FORBES, G.I., MACLEOD, A.F. and SHARP, J.C.M. 1986. The costs and benefits of preventing poultry-borne salmonellosis in Scotland. *Discussion Paper No. 05/86*. Health Econom. Res. Unit, Univ. of Aberdeen.

INDEX